National Capacity in Forestry Research

Committee on National Capacity in Forestry Research

Board on Agriculture and Natural Resources

Division on Earth and Life Studies

National Research Council

NATIONAL ACADEMY PRESS
Washington, D.C.

NATIONAL ACADEMY PRESS · 2101 Constitution Avenue, NW · Washington, D.C. 20418

NOTICE: The project that is the subject of this report was approved by the Governing Board of the National Research Council, whose members are drawn from the councils of the National Academy of Sciences, the National Academy of Engineering, and the Institute of Medicine. The members of the committee responsible for the report were chosen for their special competences and with regard for appropriate balance.

This report has been reviewed by a group other than the authors according to procedures approved by a Report Review Committee consisting of members of the National Academy of Sciences, the National Academy of Engineering, and the Institute of Medicine.

This study was supported by Contract/Grant No. 98-G-203 between the National Academy of Sciences and the Forest Service of the U.S. Department of Agriculture. Any opinions, findings, conclusions, or recommendations expressed in this publication are those of the author(s) and do not necessarily reflect the views of the organizations or agencies that provided support for the project.

International Standard Book Number 0-309-08456-3

Additional copies of this report are available from National Academy Press, 2101 Constitution Avenue, N.W., Lockbox 285, Washington, D.C. 20055; (800) 624-6242 or (202) 334-3313 (in the Washington metropolitan area); Internet, http://www.nap.edu

Printed in the United States of America
Copyright 2002 by the National Academy of Sciences. All rights reserved.

THE NATIONAL ACADEMIES

National Academy of Sciences
National Academy of Engineering
Institute of Medicine
National Research Council

The **National Academy of Sciences** is a private, nonprofit, self-perpetuating society of distinguished scholars engaged in scientific and engineering research, dedicated to the furtherance of science and technology and to their use for the general welfare. Upon the authority of the charter granted to it by the Congress in 1863, the Academy has a mandate that requires it to advise the federal government on scientific and technical matters. Dr. Bruce M. Alberts is president of the National Academy of Sciences.

The **National Academy of Engineering** was established in 1964, under the charter of the National Academy of Sciences, as a parallel organization of outstanding engineers. It is autonomous in its administration and in the selection of its members, sharing with the National Academy of Sciences the responsibility for advising the federal government. The National Academy of Engineering also sponsors engineering programs aimed at meeting national needs, encourages education and research, and recognizes the superior achievements of engineers. Dr. Wm. A. Wulf is president of the National Academy of Engineering.

The **Institute of Medicine** was established in 1970 by the National Academy of Sciences to secure the services of eminent members of appropriate professions in the examination of policy matters pertaining to the health of the public. The Institute acts under the responsibility given to the National Academy of Sciences by its congressional charter to be an adviser to the federal government and, upon its own initiative, to identify issues of medical care, research, and education. Dr. Kenneth I. Shine is president of the Institute of Medicine.

The **National Research Council** was organized by the National Academy of Sciences in 1916 to associate the broad community of science and technology with the Academy's purposes of furthering knowledge and advising the federal government. Functioning in accordance with general policies determined by the Academy, the Council has become the principal operating agency of both the National Academy of Sciences and the National Academy of Engineering in providing services to the government, the public, and the scientific and engineering communities. The Council is administered jointly by both Academies and the Institute of Medicine. Dr. Bruce M. Alberts and Dr. Wm. A. Wulf are chairman and vice chairman, respectively, of the National Research Council.

Committee on National Capacity In Forestry Research

FREDERICK W. CUBBAGE, *Chair*, North Carolina State University
PERRY J. BROWN, University of Montana
THOMAS R. CROW, University of Michigan
JOHN C. GORDON, Yale University
JOHN W. HUMKE, The Nature Conservancy, Colorado
REX B. MCCULLOUGH, Weyerhaeuser Co., Washington
RONALD R. SEDEROFF, North Carolina State University

Staff

CHARLOTTE KIRK BAER, *Study Director*
LUCYNA K. KURTYKA, *Project Officer**
KAREN BEARD, *Policy Intern*
SHIRLEY B. THATCHER, *Senior Project Assistant***
STEPHANIE PADGHAM, *Project Assistant*
NORMAN GROSSBLATT, *Editor*

*through May 2000
**through April 2000

Board On Agriculture and Natural Resources

HARLEY W. MOON, *Chair*, Iowa State University
CORNELIA B. FLORA, Iowa State University
ROBERT B. FRIDLEY, University of California
BARBARA GLENN, Federation of Animal Science Societies
W.R. (REG) GOMES, University of California
LINDA GOLODNER, National Consumers League
PERRY R. HAGENSTEIN, Institute for Forest Analysis, Planning, and Policy
GEORGE R. HALLBERG, The Cadmus Group, Inc.
CALESTOUS JUMA, Harvard University
GILBERT A. LEVEILLE, McNeil Consumer Healthcare, Denville, New Jersey
WHITNEY MACMILLAN, Cargill, Inc., Minneapolis, Minnesota
TERRY MEDLEY, DuPont Biosolutions Enterprise
WILLIAM L. OGREN, U.S. Department of Agriculture
ALICE PELL, Cornell University
NANCY J. RACHMAN, Novigen Sciences, Inc.
G. EDWARD SCHUH, University of Minnesota
BRIAN STASKAWICZ, University of California, Berkeley
JOHN W. SUTTIE, University of Wisconsin
JAMES TUMLINSON, USDA, ARS
JAMES J. ZUICHES, Washington State University

Staff

CHARLOTTE KIRK BAER, *Director*
SHIRLEY B. THATCHER, *Administrative Assistant**

*through April 2000

Preface

In the past decade, the forestry sector and the research capacity in that sector have seen substantial changes. The U.S. Department of Agriculture's (USDA) Forest Service asked the National Research Council Board on Agriculture and Natural Resources to conduct a study that focused on the nation's capacity in forestry research. Forest Service leaders recognize the necessity for improving forest productivity and stewardship of all the forests in the United States, including the national forests, urban forests, nonindustrial and industrial private forests, and tribal, state, and community forests. Continuous research findings must inform the management and protection of the forests. However, our national capacity in forestry research appears to have waned even as the demands placed on our forests and the need for enhanced technical knowledge have increased. We must have better information on the status of forestry research and future research priorities if we are to identify critical research needs and we need to identify the types of scientists and disciplines required to produce knowledge about our nation's forests.

This study of our nation's capacity in forestry research was conducted to review the expertise and future needs of forestry-research organizations and to review the current approaches and capacity of natural-resource education to address shortfalls of scientists expected in selected disciplines in the next 10 to 15 years. In performing our assessment, we relied on a wealth of background information about forestry research and education capacity. We obtained literature, policy statements, strategic plans, and white papers from many organizations interested in forestry research and education. We sponsored a workshop on forestry-research capacity on July 16–17, 1999, which included speakers, focus groups, and comments from interested organizations. The workshop was an important component of this project because it provided direction to the report. During the workshop, participants were asked to address questions that were part of this committee's task in breakout sessions. First, the participants were asked to determine critical issues and priorities in forestry. Then, they were asked to determine how these priorities should be met in relation to knowledge base, research capacity, interdisciplinary and spatial applications and incentives, and university curricula and programs. The input that the committee received from the workshop participants was recorded and used to direct the study and recommendations presented in Chapters 2 through 5. We also collected background material on budgets, scientific efforts, and trends in graduate education.

Clearly, the USDA Forest Service remains the largest forestry research organization in the world, but has experienced fairly steady declines in real funding levels and in personnel and facilities. Other federal government organizations, such as the National Aeronautics and Space Administration (NASA), the National Science Foundation (NSF), the Department of Energy (DOE), and the Environmental Protection Agency (EPA), have increased their broad focus on terrestrial research and development, in either applied or basic sciences. Universities provide almost as much support for forestry research as the Forest Service. They have slightly fewer scientist full-time equivalents (FTEs) dedicated to research than does the Forest Service, but more than double their FTEs in teaching and extension. The forest products industry performs some forestry research directly and contributes some funds to university and other research cooperatives. In total, annual expenditures on forestry research probably exceed $500 million per year. About 1400 scientist FTEs are dedicated to forestry research in the United States, as well as 600 teaching FTEs and 240 extension FTEs.

The significant amount of resources expended on forestry research and the substantial number of scientists working in these areas provide capacity for analyses of many subjects. The diversity of funding sources and organizations involved in forestry research provides avenues for incorporating different perspectives on the multiplicity of important forest values and some competition among agencies to provide leadership in areas uniquely related to their mission. The USDA Forest Service has focused on traditional forest management and protection questions, and has expanded their purview to emerging issues such as sustainable forest management, global change, and forest monitoring. The National Research Initiative within USDA has focused on more basic biological forestry science; NASA on remote sensing applications of ecological issues; DOE on industrial energy or competitiveness; EPA on terrestrial impacts on air or water systems; and the U.S. Geological Survey and Fish and Wildlife Service on terrestrial and aquatic fauna.

Of these organizations, the USDA Forest Service has experienced slightly decreased research funding and capacity in terms of real dollars, and the forest products industry probably has as well. Universities have had stable personnel numbers in total, but dynamic fluctuations at individual institutions. The other organizations and sectors are relatively new contributors to the nation's forestry research capacity and expertise. Thus one's perception of problems in forestry research capacity depends on one's perspective. The addition of capacity from new organizations is welcome, and indeed should be augmented if possible. The reduction of capacity in the USDA Forest Service is cause for concern. These trends outlined in this report should promote agency introspection about the direction of and support for their programs, and serious collaboration with external clients to redress the causes as well as symptoms of that decline.

Despite the diversity and relative depth of forestry research capacity, this report identifies critical needs and makes suggestions for significant improvements. In the committee's opinion forestry research capacity is at a crossroads, if not a precipice. First and foremost, the population in the United States and the world continues to increase moderately, while the forest area is stable at best, if not actually declining. Furthermore, the number of demands for commodity production and for environmental services from forests has increased at least as rapidly as population, and perhaps faster as forests become fewer or more fragmented. Even successful management efforts to produce more commodities (e.g., timber) or services (e.g., recreation) must meet much stricter environmental and social standards than in the past, which may impede short-run productivity gains. New factors are affecting forestry, such as the international agreements on Sustainable Forest Management and on Criteria and Indicators, large increases in forest recreation and tourism, demands for water quality protection and use of Best Management Practices, concern about invasive species and fire, advances in biotechnology, and broad based regional assessments. Political debates about forestry issues at the local, state, national, and world levels have increased greatly, but government funding, legislation, and reform often faces gridlock. On the other hand, market forces such as certification and international competition for market share have made major changes in corporate forestry ownership, structure, and practices.

This dichotomy between more demands for forest products and services and fewer forests suggests that we need more intensive management of some areas for timber and commodities; more areas set aside or managed for wildlife, biodiversity, recreation and non-market values; and more cooperation among various stakeholders on public and, increasingly, on private forest lands. Better research and implementation of those research advances provide the only practical means that we can meet increased demands with decreased supply.

Will we be able to satisfy these increased forestry research needs? Have our efforts to date been satisfactory? Do we have adequate scientists, facilities, management, and support for research efforts now? Will the forestry research capacity be prepared for the likely future emerging issues? Will the current status quo suffice? In brief, this report suggests that our current forestry research capacity is neither adequate now, nor poised for success in the coming years. This report identifies significant declines in real research capacity, fragmented cooperation and poor communication among the principal providers and users of forestry research, inadequate support of both foundation and emerging disciplines, and little strategic planning to address future forestry research needs.

The forestry research sector is indeed at a crossroads. If left unchanged, its future will entail a steady erosion of intellectual and institutional capacity, and dwindling capacity and impact. Alternatively, forestry research could renew its commitment to innovation, cooperation, relevance, and extension in order to prosper and enhance the practice of forestry in this century. This latter vision will require levels of cooperation, support, real exchange of financial and technical support, and stakeholder support that do not currently exist. This report of forestry research capacity makes recommendations that will help achieve this positive, proactive role for forestry research in the future. It summarizes our findings and recommendations regarding each of those components of our assessment of national forestry-research capacity. It presents our conclusions about the status of forestry-research capacity and our specific recommendations for enhancing it. Our review and our recommendations can be used to shape future forestry-research efforts, enhance research capacity, and encourage public and private interests to help to achieve a strong research foundation for sustainable forest management.

Forestry research has many strengths and beneficial effects. A progressive and productive relationship among all the key players in forestry research is essential. We hope that this report will be useful to those players, including federal and state entities, university and research organizations, industry and business, student populations, and those in positions of decision-making that will affect future generations.

FREDERICK W. CUBBAGE
Chair, Committee on
National Capacity in Forestry Research

Acknowledgments

The Committee wishes to thank the many people who provided input by letter or at the public workshop. Special thanks are due to Richard Guldin, U.S. Department of Agriculture Forest Service, for his willingness to respond to the Committee's numerous requests for data and information. Without his assistance and perseverance, the report would not have become a reality.

This report has been reviewed in draft form by individuals chosen for their diverse perspectives and technical expertise, in accordance with procedures approved by the National Research Council's (NRC) Report Review Committee. The purpose of this independent review is to provide candid and critical comments that will assist the institution in making its published report as sound as possible and to ensure that the report meets institutional standards for objectivity, evidence, and responsiveness to the study charge. The review comments and draft manuscript remain confidential to protect the integrity of the deliberative process. We wish to thank the following individuals for their review of this report:

Jo Ellen Force, University of Idaho
Perry Hagenstein, Institute for Forest Analysis, Planning, and Policy
John A. Helms, University of California, Berkeley
T. Kent Kirk, Verona, WI
Dennis LeMaster, Purdue University
Kenneth Munson, International Paper
John Pait, The Timber Company
Paula Stephan, Georgia State University

The review of this report was overseen by Ellis Cowling, North Carolina State University, and Henry Riecken, University of Pennsylvania. Appointed by the National Research Council, they were responsible for making certain that an independent examination of this report was carried out in accordance with institutional procedures and that all review comments were carefully considered. Responsibility for the final content of this report rests entirely with the authoring committee and the institution.

Finally, the committee wishes to thank Charlotte Kirk Baer, study director, for her encouragement and guidance of this project to completion.

Contents

Executive Summary, 1
    Defining Forestry-Research Capacity, 2
    The Value of Forestry Research, 2
    Key Players, 3
    Knowledge Base and Priorities, 3
    Assessing the Status of Forestry Research, 4
    Enhancing Forestry-Research Personnel, Facilities, and Infrastructure, 5
    Leadership and Strategic Planning, 6
    Creating Intellectual and Scientific-Research Capital, 9
    Increasing Strength, Collaboration, and Diversification in Forestry Research, 10
    Ensuring Progress, 12

1. Need, Context, and Foundation for Forestry Research, 14
    The Current Study, 15
        Boundaries of the Assessment, 16
    Defining Forestry-Research Capacity, 16
    Institutional Framework for Forestry-Related Research, 17
    Early Forestry Research and Education, 18
    The Importance of Maintaining, Protecting, and Enhancing Today's
    Forests for Tomorrow, 19
    Future Challenges, 21
    Forestry Education and Research, 21

2. The Essential Knowledge Base for Forestry Issues, 23
    Knowledge Base Required, 25
        Foundation Education and Research Priorities, 26
        Emerging Education and Research Priorities, 26
        Stewardship and Sustainability of Public Lands, 28
        Sustainable-Management Criteria and Indicators, 28
        Forest Certification, 33
        Forest-Industry Priorities, 36

New Forestry-Research Challenges, 37
Workshop Input on an Essential Knowledge Base, 37
Conclusions and Recommendations, 40

3. Current Forestry-Research Capacity in the United States, 42
    Assessing Forestry-Research Capacity, 43
    A Portrait of the Forestry-Research Workforce, 43
        USDA Forest Service, 44
            Research Scientists, 47
            Research Productivity, 49
            Research Quality, 51
            Research Advisory Body, 51
        Professional Forestry Schools and Colleges, 52
            Faculty, 52
            Forestry Extension, 53
        Private Industry, 56
        Total Forestry Research Workforce by Sector, Function, and
            Sustainable Forest Management Criteria, 56
    Investments in Forestry Research, 59
        Forest Service Research Support, 59
        Other Federal Forestry-Research Funding, 61
        Leveraging Research Support, 63
        University Research Support, 68
        Contributions of the Forest Products Industry, 68
        Other Sources of Research Support, 70
    Evaluating Return on Investment in Forestry Research, 73
    Conclusions and Recommendations, 76
        Personnel, 76
        Research Quality, Productivity, and Efficacy, 78
        Fiscal Strength, 80
        Toward Greater Capacity, 81

4. Preparing Forestry Scientists and Users of Forestry Science, 82
    The Future of Forestry Education, 83
    Trends in Enrollment and Graduation, 85
    Forestry as an Academic Subject, 87
    Curriculum as a Concept, 87
    Models for Forestry Education, 88

CONTENTS   xv

    Broad Trends in Forestry Education, 90
  What About Research?, 91
  What about Curricula?, 92
  Adequacy and Capacity of University Programs
  to Meet Near-Future Needs, 93
    Disciplinary Breadth of Forestry Education, 93
    Numbers of Scientists, 94
    Diversity of Scientists, 98
    Future Demand for Scientists, 99
  Interdisciplinary and Integrative Capabilities, 100
    Institutional Arrangements, 100
  Conclusions and Recommendations, 102

5. Capacity of Forestry-Research Organizations to Meet Future Research Needs, 105
  Continuity Through Time: Resources to Maintain Operations, 106
    University System, 107
    Forest Industry, 110
    USDA Forest Service, 111
  Facilities and Equipment to Perform High-Quality Research, 112
  Access to People with Appropriate Skills and Competences, 114
  Focus on High-Priority Goals and Needs, 115
  Conclusions and Recommendations, 116

6. Summary and Conclusions, 119
  Recommendations, 120
  Conclusions, 127

References, 128

Appendixes, 136
  Appendix A—Workshop Agenda, 136
  Appendix B—Breakout Group Questions, 139

About the Authors, 140

Tables and Boxes

Tables

1-1 Definitions of terms commonly used to describe what forests provide, 20
2-1 Comparative forestry-science education and research priorities according to selected sources, 27
2-2 Criteria for the conservation and sustainable management of temperate and boreal forests developed from the Montreal process, 30
3-1 Forestry-research statistics for the USDA Forest Service, fiscal years 1980-1999, 45
3-2 USDA Forest Service research funding by budget line item, fiscal years 1980-1999, 46
3-3 Number of Forest Service research scientists by discipline, fiscal year 1985-1998, 48
3-4 Number of Forest Service publications by discipline, fiscal years 1981-1998, 50
3-5 Trends in forestry employment in universities, 54
3-6 Full Time Equivalents of U.S. Forestry Scientists by Sector, Function, and SFM Criterion, 2001, 58
3-7 McIntire-Stennis funding in actual and constant dollars, fiscal year 1980-2000, 64
3-8 Distribution of McIntire-Stennis funds to eligible state institutions of institutional units fiscal year 2000, 65
3-9 Sustainable-forestry research funding by industry through the Sustainable Forestry Initiative program, in dollars, 69
3-10 Federal funding for forestry research by selected agency and program, fiscal years 1994-2000, 71
3-11 Returns on investments in forestry research, 75
4-1 Enrollment and degrees awarded in forest science programs, 1989-1998, 86
4-2 Enrollment in forestry, natural resources, and agriculture programs by program and degree level, 1993-1999, 86
4-3 Enrollment in forest science programs by academic specialization, 1993-1999, 95
4-4 Forest sciences enrollment statistics by gender, ethnicity, and citizenship, fall 1999, 98

Boxes

2-1 Excerpts from input received on education and research needs to form an essential knowledge base, 38
3-1 Hatch and McIntire-Stennis proposals at the College of Agriculture and Life Sciences, University of Wisconsin, 62
3-2 Reviews improve quality of forestry research at the University of California, Berkeley, 67

4-1 Graduate student support, 97
4-2 The corporate environmental management program at the University of Michigan - an example of creative partnerships within the University, and between business and the University, 100
4-3 National Science Foundation's Integrative Graduate Education and Research Training program, 101
4-4 NSF's Luquillo Long-Term Ecological Research (LTER) – An Example of Forestry Research Conducted through a creative partnership between universities and federal research agencies, 102
5-1 Northwest stand management cooperative, 106
5-2 Centers of excellence in forestry, 107
5-3 Virtual center concept at work, 109

Executive Summary

Forests are major components of the earth's natural resources and they are increasingly critical to the welfare of the U.S. economy, environment, and population. Desires to improve forest management and productivity, preserve biodiversity, maintain ecologic integrity, and provide societal services, such as recreation and tourism, necessitate a strong forestry-research base.

Given the clear importance of forestry research in sustaining forests for the future, the U.S. Department of Agriculture (USDA) Forest Service asked the Board on Agriculture and Natural Resources of the National Academies to undertake a study of the nation's capacity in forestry research. The Committee on National Capacity in Forestry Research was appointed to carry out the study, which was conducted to review the current expertise and status of forestry research and to examine the approaches of natural resources education and forestry-research organizations to meet future needs. The committee was charged with the tasks of assessing

- The knowledge base necessary for forestry experts and other professionals to address research and management issues successfully in a complex social, political, and technical environment.
- The capacity of research organizations that employ those professionals to perform research that will provide the basis of scientific management and protection of the nation's forest resources.
- The basic curriculum elements and level of instruction necessary to develop a core competence, requisite to the desired knowledge base, that will produce suitably trained, socially aware, and technically proficient researchers and managers.
- The means by which focused education and interdisciplinary systems thinking and communication skills can be developed and applied by a wide array of professionals to forest-landscape problems.
- The adequacy and capacity of available university-level programs to meet near future needs.

Our analysis and recommendations place special emphasis on the nation's largest forestry research entity, the USDA Forest Service, but they also address its larger operating environment, including other public and private research and educational institutions that supply forestry-research capacity.

A major source of input for this study was a workshop that took place on July 15–16, 1999, at the National Academies in Washington, D.C. In addition to the workshop and associated public comments and letters, the committee communicated with professionals in relevant forestry research and education organizations and consulted numerous other sources, including recent surveys and studies of trends in forestry education, to obtain relevant information for analysis.

DEFINING FORESTRY-RESEARCH CAPACITY

Assessing national capacity in forestry research requires definitions of the general scientific concepts of forestry and of the notion of research capacity. An understanding of return on investment in forestry research is also needed.

For purposes of this study, a modified definition of forestry is adopted from definitions published by several sources.

Forestry is the science, art, and practice of creating, managing, using, and conserving forests and associated resources in a sustainable manner, engaging broad and specialized scientific disciplines to meet desired goals, needs, and values.

A comprehensive definition of research capacity is difficult to capture because it has no fixed boundaries. Capacity encompasses human resources, institutions, infrastructure, and financial support.

Research capacity is the magnitude of the ability to develop, advance, and disseminate science and technology.

THE VALUE OF FORESTRY RESEARCH

The estimated return on investment in wood-product and timber-management research has been reported to be as high as 40 to 86 percent per year. Forest products and use research conducted by the Forest Service, for example, has contributed to the development of knowledge and technology that have tripled the amount of fiber available for use from trees within the last 100 years, greatly extending forest resources. Research on recycling of wood-based products has increased paper-recovery rates from 25 percent to 45 percent of fiber. A specific example is the scientific advance in recycling of 33 billion stamps produced each year by the U.S. Postal Service as a result of research on pressure-sensitive adhesives, which had presented substantial problems in recycling. Other research advances include the development of composite products and improvement in housing constructions. Similarly, research conducted by universities, industry, and government on forest health, genetics, management (intensive, extensive,

EXECUTIVE SUMMARY

and alternative), and fire has contributed to improved planting stock, tree growth and quality, and forest sustainability.

The value of forestry research has been measured by the gains that have increased the efficiency of wood use and timber management, but these measures do not address the gains attributed to productivity research or the benefits derived outside the marketplace, such as those related to environmental protection and social welfare. Although it is not remembered well, the Forest Service was established to protect watersheds and maintain the nation's supply of fresh drinking water, and watershed research has retained high priority. Research contributions include assessing effects of National Forest and other owners' land-management activities on drinking-water source quality; these assessments are required by the Environmental Protection Agency (EPA) to determine the risk of contamination of drinking-water sources and to enable science-based decision-making. Human dimensions of natural resources management go beyond health to encompass social aspects. Research by social scientists on how human behavior, social institutions, demographic needs, and values affect the availability, demand, and use of forests is increasingly important to sustainability of these lands.

KEY PLAYERS

The analysis of national research capacity in this report is focused on major forestry-research organizations that are the key players in terms of capacity. Those organizations include the Forest Service and several other federal agencies that conduct forestry-related natural resources research (for example, the Department of Energy, DOE; EPA; the National Aeronautics and Space Administration, NASA; the Department of Interior, DOI; and the National Science Foundation, NSF), nongovernment organizations, professional forestry schools and colleges, other university departments and research units, and the private forest industry. Our analyses and recommendations to enhance the nation's forestry-research capacity are related to those players.

KNOWLEDGE BASE AND PRIORITIES

Assessing the knowledge base in forest sciences and the nation's research capacity in forestry involves the identification of current knowledge gaps. Forecasting future research needs and capacity for improved forest management, protection, and production requires the identification of education and research priorities to fill the gaps and to support current endeavors. Traditional areas of science provide the foundation for all work in forestry. These foundation fields of science education and research include biology, ecology, and silviculture; genetics; forest management, economics, and policy; and wood and materials science. Numerous gaps in knowledge related to various specific scientific aspects of forestry have been identified, but there is general agreement among forest researchers that basic biologic knowledge is limited and that our understanding of forest health, systems, and management and wood science is deficient. Because of new and evolving roles of forests in our society and because of changing emphasis in and goals of forest management and protection, several subjects of education and research will be increasingly important in the future; these emerging subjects include human-natural resource interactions; ecosystem function, health, and management; forest systems in various scales of space and time; and forest monitoring, analysis, and adaptive management, and forest biotechnology.

Recommendation 2-1

To achieve an adequate knowledge base, forestry and natural-resource education and research programs in government and academia should dedicate resources to the foundation fields of forestry science while engaging in efforts to develop emerging education and research priority areas.

ASSESSING THE STATUS OF FORESTRY RESEARCH

In obtaining information for this report, we were challenged by the limited availability of systematic budgetary, expenditure, and programmatic data on the diverse forestry and natural resources programs from the different agencies and organizations responsible for or involved in research. The Forest Service has taken the lead in systematically compiling and tracking that type of information. Although the Forest Service maintains pertinent information related to its research activities, there is a lack of comprehensive information on forestry research in the United States.

In 1997, the National Science and Technology Council recommended a framework for integrating the nation's environmental monitoring and research networks and programs, noting that new developments in science and technology provide new opportunities for collecting and organizing data. With current fiscal limitations facing all levels of government, cooperation and efficiency among agencies is essential to the long-term success of individual programs. Following on the need for an integrated environmental and monitoring network, an integrated research-information system is needed for tracking forestry research activities. The initial challenge will be to build on, enhance, and integrate existing databases.

Recommendation 3-1

The Forest Service should enhance its current research-information system and tracking efforts by establishing an improved and integrated interagency system that includes relevant information on forestry research activities, workforce, funding, and accomplishments in all agencies of the U.S. Department of Agriculture, other relevant federal agencies, and associated organizations as appropriate.

Integrating information on forestry research from the Forest Service with information from other agencies in USDA and for example, from DOE, EPA, DOI, NSF, and NASA will provide a stronger foundation on which to base decisions. Developing better information on the status of forestry research will require settling on the types of data to include in such a system; determining funding and staffing levels of federal, state, university, and nongovernment organizations that perform forestry research; noting research priorities; and tracking quantitative and qualitative research accomplishments. Such a strategy is essential to monitoring the nation's research capacity and to differentiate between actual and perceived advances in forestry research.

ENHANCING FORESTRY-RESEARCH PERSONNEL, FACILITIES, AND INFRASTRUCTURE

Scientific discovery and productivity can depend as much on advances in specific scientific disciplines as on the personnel, facilities, and infrastructure through which they are founded. Concerns expressed by members of the scientific-research community concentrate on the decreasing number of researchers and waning attention to research facilities and infrastructure. Indeed, the USDA Forest Service, the world's largest forestry-research organization, has experienced a 46 percent decrease in number of scientists in the last 15 years, from 985 in 1985 to 537 in 1999. It is clear that Forest Service research capacity has decreased dramatically in terms of numbers of scientists. Despite apparent and measured increases in efficiency, effectiveness, and productivity, the Forest Service research base has dwindled as population demands on our forest resources have increased.

Recommendation 3-2

The Forest Service should substantially strengthen its research workforce over the next five years to address current and impending shortfalls, specifically recruiting and retaining researchers trained in disciplines identified as foundation and critical emerging fields of scientific education and research.

Strengthening the Forest Service's ability to respond to short- and long-term research needs is essential. In an attempt to account for recent shortages of research scientists in specific fields, the Forest Service has routinely supplemented its workforce with temporary employees to work on critical issues. However this approach does not lend itself to continuity or the ability to adequately address research priorities. Employing additional full-time permanent researchers, rather than supplementing with temporary employees and post-doctoral students in fields that are required to address traditional and emerging issues, will improve Forest Service continuity and effectiveness in research efforts, especially long-term projects. Deficiencies in the forestry research workforce should be addressed as soon as possible, because trends to date indicate that the situation may worsen. In the past 8 years the Forest Service lost more than 9000 total employees and during the past 15 years has lost approximately 45% of its scientists. Currently 35% of its workforce is eligible to retire in the next five years and the average age of employees is 55 years. The cost associated with strengthening and retaining the Forest Service research workforce is nominal compared with the costs associated with operating under current and projected deficiencies.

Recommendation 3-3

As part of the increase in research personnel capacity and resources, the Forest Service should enhance cooperative relations with forestry schools and colleges.

The National Science and Technology Council states that partnerships between the federal government and the nation's universities have proven exceptionally productive, successfully promoting discovery of knowledge, stimulating technologic innovations, improving quality of life, educating and training the next generation of scientists and engineers, and contributing to America's prosperity. However, cooperative research allocations as a proportion of the Forest Service research budget have decreased markedly from about 15 percent to 9 percent in the last seven years (see Table 3-1). Given an environment of decreasing budgets and fiscal constraints, the Forest Service should consider allocating a larger portion of its total research budget to the station or research work unit level for extramural research grants that are inter-organizational and cooperative, requiring active involvement, cooperation, and integration of Forest Service, university, and other research partners. Two important rationales exist for federal investment in university-based research: (1) the benefits derived from training a new generation of scientists and (2) continuous mutual scientific and financial enrichment that is derived from the relationship.

LEADERSHIP AND STRATEGIC PLANNING

The forestry-research sector consists of a broad group of public and private organizations. As the largest forestry-research organization in the world, the USDA Forest Service must provide research leadership. Strong leadership is accomplished by defining a clear vision for research and communicating effectively with research interest groups. Successful strategic planning is accomplished by setting goals and measuring progress toward them.

The current scarcity of resources calls for improved collaboration, communication, and oversight of forestry research. A central organizing body is needed to help set research priorities, oversee monitoring of research accomplishments, and facilitate cooperation among research organizations. Creation of new federal or state organizations is not necessary, but the roles of existing forestry-research oversight bodies should be refined and improved.

EXECUTIVE SUMMARY

Recommendation 3-4

The USDA Forest Research Advisory Committee should focus its efforts in two primary areas: (1) work with USDA research leaders in the Forest Service and other agencies to set research priorities and monitor accomplishments, and (2) coordinate with USDA's Cooperative State Research, Education, and Extension Service and other agencies to help guide research priorities of McIntire-Stennis, Renewable Resources Extension Act, National Research Initiative, and other grant programs.

Currently, the Forest Research Advisory Council is tasked to provide advice to the Secretary of Agriculture on forestry issues and accomplishing the purposes of the McIntire-Stennis Act. Its membership is currently drawn from state and federal government (USDA and EPA), university, industry, and nongovernmental organizations. Staff support to the group is 0.3 staff year equivalent and the group meets at least once per year. To be effective, the advisory body should include professionals in several other government agencies, a broader spectrum of universities, and relevant organizations as regular or exofficio members. A full-time dedicated professional USDA senior-level director would facilitate operations, serve as communication liaison, collect and monitor data on forestry-research accomplishments, and coordinate site reviews and visits. The advisory body and staff would also monitor forestry-research quality and accountability by renewing and expanding the periodic-review process, including reviews of McIntire-Stennis projects and Forest Service research accomplishments. Reasonable intervals for site visits are 10 years for McIntire-Stennis institutions and 5 years for Forest Service research stations.

A more focused advisory committee would help to ensure that research agencies and organizations are pursuing appropriate strategic directions and implementing them with sound operational programs. Implementing or renewing forestry-research oversight reviews would correspond with the mandates for performance evaluation under the Government Performance Results Act (GPRA). External peer reviews and funding competition would encourage increased consistency and higher quality of formula-funded research. All these steps would foster better communication about programs and support for their missions.

Traditional programs to support forestry research and education will be better served by focused guidance. Those programs include the McIntire-Stennis program and the Renewable Resources Extension Act (RREA).

Recommendation 3-5

Universities and state institutions should increase the use of competitive mechanisms for allocating McIntire-Stennis and Renewable Resources Extension Act funds within these institutions, and in doing so, encourage team approaches to solving forestry and natural resource problems as well as integrated research and extension proposals or interinstitutional cooperation.

With goals consistent to the respective Congressional Acts, universities can allocate McIntire-Stennis and Renewable Resources Extension Act (RREA) funding via a merit-based competitive process. Scientific excellence is recognized to be promoted when investments are guided by merit review that rewards quality and productivity in research and accommodates for endeavors that might be high-risk but have potential for high gain.

Clearly, formula-funds such as McIntire-Stennis are critical for diffusing research throughout the nation, for pursuit of long-term research goals and multidisciplinary research, and for supporting a system in which university faculty appointments are split among some combination of research, extension and teaching. There is a need to preserve the advantages offered by formula funding, particularly their facilitation of linked research, extension, and teaching programs. However, if more competitive approaches were used by universities and state institutions for allocation of formula-based McIntire-Stennis funds, the opportunities for improving the quality and accountability of research funded will be greater. A stronger commitment to addressing the quality and accountability of formula-based research might garner greater support for funding the critical McIntire-Stennis program at a level closer to that at which it was authorized. The current funding level of McIntire-Stennis is only approximately $21 million, which is less than half its authorized level. In light of this limited funding, institutions should concentrate research capital in specific (and perhaps limited) fields of forestry research where they operate best or have some recognized institutional advantage.

In addition to research oversight and mechanisms, technology transfer should be improved. We have made continuous strides in many fields of basic and applied research, but real resources directed to extension and cooperative efforts have steadily declined. A stronger delivery system must be developed.

Recommendation 3-6

The U.S. Department of Agriculture, together with universities, should develop means to more effectively communicate existing and new knowledge to users, managers, and planners in forestry.

The United States has almost 10 million nonindustrial private forest landowners, who own 49 percent of the nation's forest land and 58 percent of the nation's commercial timberland. Forestry and natural-resources extension programs provide direct support for disseminating research findings to research users, such as nonindustrial private forest landowners, urban residents, production and environmental interest groups, natural-resource professionals, state and federal agencies, local governments, and policy-makers.

The USDA maintains a unique position to communicate research results to everyday users through its extension programs. That capacity does not all need to be housed at or be used by the Forest Service research branch, as suggested in a related study by the Strategic Planning Task Force on USDA Research Facilities. Without effective mechanisms for technology transfer and adoption, forestry-research findings become a mere collection of observations and data. Forestry and natural resources extension programs have played a major role in communicating research results to users. To strengthen that role and ensure continuity in technology transfer, universities, government, and private organizations should actively participate in training forestry researchers to communicate research results to forest managers and to be receptive to their needs.

CREATING INTELLECTUAL AND SCIENTIFIC-RESEARCH CAPITAL

Despite constraints on growth in the forest sciences, colleges and universities must develop the next generation of scientific leadership. That requires depth and breadth of undergraduate and graduate education and experiences. Undergraduate programs should educate students in the basic forest sciences, and opportunities for specialization or diversification should be encouraged at the graduate level.

Focused education in basic science-including field biology, population genetics, plant systematics, and plant taxonomy is fundamental to understanding any biologic system. Declines in fundamental disciplines—such as genetics, physiology, pathology, and entomology—have been observed in faculty and support staff of universities and natural resources agencies. Demands for social-science knowledge have increased greatly, but scientific staff in this area remains at historically low levels. The intellectual capital in many of these fundamental areas is dangerously low, and this lack of capacity will affect the nation's ability to implement new programs of research and development. The challenge is to find the means by which truly focused education and interdisciplinary systems thinking and communication skills can be developed and applied by forestry professionals.

Recommendation 4-1

University programs should assume a renewed commitment to the fundamental areas of scholarship and research in forest sciences that have diminished in recent years, and adopt an enhanced, broad, integrative, and interdisciplinary programmatic approach to curricula at the graduate level.

The next generation of forestry researchers will require skills in oral and written communication, interpersonal relations, and problem-solving; fundamental and specialized knowledge in a scientific discipline; the ability to operate in a team setting; and the ability to address complex forestry and natural resources research challenges.

In addition to formal "systems" courses, such as ecology, "systems thinking" should be embodied in teaching and learning through the use of examples in which the description and integration of systems components are demonstrated. The systems approach can be enhanced if all future researchers have a core foundation in scientific method and discipline and have a specialization in which they have a competent depth and an appreciation of a wide range of other disciplines, including the ability to communicate effectively with scientists in other disciplines. Breadth and depth are both essential in graduate programs.

Managers of academic programs must be aware of the time and resources necessary to support synthesis and cooperative efforts by faculty and students. Promoting and achieving disciplinary integration is difficult but not impossible. Creative approaches involve the study of natural resource issues and the use of core and capstone courses that blend biological and social sciences.

In interdisciplinary and interinstitutional efforts, scientists must be trained not only in a technical skill, but also in skills that allow them to work in complex teams focused on common goals. The present reward systems tend to work against cooperative models by favoring and rewarding individuals. A system that encourages both individuals and teams without stifling individual creativity should be developed.

Recommendation 4-2

Universities should develop joint programming in geographic regions to ensure a "critical mass" of faculty and mentoring expertise in fields where expertise might be dispersed among the universities.

Because there is a wide variety of subfields in forestry and natural resources and few institutions can produce doctoral graduates in many subfields, regional cooperation might be viewed as a way to expand capacity by pooling resources in important areas. Building of regional coalitions among universities for the purpose of graduate education could enhance the education of students and lead to cost-effective expansion of the capacity to develop forestry and natural resources scientists.

INCREASING STRENGTH, COLLABORATION, AND DIVERSIFICATION IN FORESTRY RESEARCH

Most persons agree that the forestry research enterprise must do more research with fewer resources, collaborate more on projects of mutual interests, and take a broader perspective in research conducted. The nation's current research structures were based on decades of incremental improvement, and recommendations provided here do not suggest casting these structures aside as much as modifying them.

Current research organizations have merits, but we need to move toward new systems appropriate for new social and political environments. Existing resource management organizations must cooperate better, and partnerships that improve on unilateral research possible by single organizations must be formed. Research

EXECUTIVE SUMMARY

cooperatives and research consortia are one evolving means of developing research synergies. Research cooperatives and consortia provide a means for cooperation among partners—universities, industry, and states, federal, and nongovernment organizations.

Universities, government, industry, and private groups can partner to a much greater extent than in the past to ensure that the entire spectrum of forestry research and development interests is addressed and to ensure that limited resources are utilized to best advantage. Creation of centers focused on specific research emphasis that involve many players is an increasing need as forestry research continues to broaden and demands continue to expand.

Several successful examples of federal programs represent innovative approaches to education and research and foster collaboration and diversification (see Chapter 5). Those programs are examples of programs that could be implemented by USDA to improve disciplinary and multidisciplinary forestry education and research. One example is the National Science Foundation (NSF) Long Term Ecological Research Network.

Recommendation 5-1

Centers of excellence in forestry, should be established and administered by USDA. These programs and awarded projects should (1) support interdisciplinary and interorganizational activities, (2) focus on increasing underrepresented student participation in education and research, (3) clearly justify how new forestry research approaches and capacity will be enhanced, and (4) undergo initial and periodic review.

Establishing centers of excellence in forestry for fields related to forestry research and education will require investment. The magnitude of investment will depend on the type of centers established. As noted by the National Research Council in 1990, centers need not be "bricks and mortar." Options for "virtual" centers described in the current report address the need to work within the existing structure and fiscal constraints. Regardless of the type of center established, focusing research efforts and increasing efficiency of existing resources through centers will result in enhanced research and education. The goals of centers of excellence in would include: (1) working closely with government agencies and other organizations to develop new research and education collaborations and partnerships; (2) encouraging and providing opportunities for university faculty and government researchers to conduct integrated interinstitutional research; (3) providing incentives for minority group students to enter and remain in forestry research; (4) establishing measurable program goals and objectives; and (5) developing and implementing evaluations to assess the effectiveness and outcomes of programs and financial performance.

Effective recruitment and outreach by universities and governments are essential for reaching all sectors of society. However, forestry education and research have been largely ineffective in those respects over the last several decades. Minority-group participation in science education, graduate-level training, and forestry teaching, research, and development is inadequate. Recruitment and outreach need greater attention and resources. Although they should enhance minority-group participation in forestry research, a portion of it should also be targeted specifically at topics identified in Chapter 2 where there is considerable need. Achieving a more ethnically and racially diverse group of forestry scientists will require extraordinary efforts to recruit and encourage minority-groups students to pursue science careers in forestry and natural resources.

Supporting such students through awards and provided through centers of excellence in forestry is one key factor in ensuring a better prepared and more diverse workforce in the future.

Recommendation 5-2

Clear federal research facility mandates—such as long-term ecological research sites, experimental forest and natural resource areas, and watershed monitoring facilities—should receive priority for retention and enhancement, and a system of periodic review of all facilities should be implemented and maintained.

Possibilities for co-locating, virtual research centers, centers of excellence in forestry, or other collaborative research centers should be pursued for future federal forestry-research projects. Funding for future research centers should require clear justification based on the criteria of the three classes for federal research facilities (uniquely federal, appropriately federal, and not uniquely or appropriately federal). Current funding levels for facility maintenance and operation should increase at the rate of inflation to ensure a sound infrastructure.

The Agricultural Research, Extension, and Education Reform Act of 1998 states: "The Secretary shall continue to review periodically each operating agricultural research facility constructed in whole or in part with federal funds, and each planned agricultural research facility proposed to be constructed in whole or in part with federal funds, pursuant to criteria established by the Secretary to ensure that a comprehensive research capacity is maintained." Review is the only means of ensuring that objectives are being met. As previously recommended by several expert panels, the Forest Service, universities, and other forestry-research partners should review research facilities and determine how to optimize research infrastructure.

ENSURING PROGRESS

Taking into account budget limitations and the need for clearly focused programs, the recommendations offered in this report suggest areas for improvement. To ensure enhanced forestry-research capacity, we must implement principles for strategic planning to accomplish research goals, establish tracking and accounting as management and decision-making tools in research and development programs, develop innovative and contemporary models for education and research programs and infrastructure, and increase collaboration and diversification. Focusing on improvements along those lines and implementing the specific recommendations will improve existing research and education efforts, direct resources to critical needs, and provide for a healthy and vigorous forestry-research base to address future challenges.

The next step for progress to be made will require implementation of detailed suggestions contained in this report. It will take a cooperative effort of university, federal, state, and private research interest groups to actively pursue the means to implement these recommendations. The analyses summarized here and the concomitant recommendations will enhance the nation's forestry-research capacity. Follow through is required to ensure interorganizational cooperation, adequate funding, administrative tracking, educational excellence, and, most important, strong research capacity. The future of forests and of their capacity to play their accustomed roles in natural resources and social landscapes throughout the world will depend on national ability to: develop better knowledge; use that knowledge to address issues of economic, environmental, and social importance; deliver the knowledge to forest landowners and managers; and measure and monitor our progress toward achieving our universal goals.

1

Need, Context, and Foundation for Forestry Research

Since the National Research Council published its last assessment of forestry-research needs in 1990, forestry research has continued to contribute to the lives of millions in public and private sectors throughout the world (National Research Council, 1990). Forestry research has continued to enhance the health and productivity of forest resources that people have depended on and will depend on for centuries. In this report, the Committee on National Capacity in Forestry Research presents a review of the current expertise and status of forestry research and examines the approaches of natural resource education and forestry-research organizations to meet future needs.

The committee was charged with assessing

1) the knowledge base necessary for forestry experts and other professionals to address research and management issues successfully in a complex social, political, and technical environment;
2) the capacity of research organizations that employ those professionals to perform research that will yield a basis for scientific management and protection of the nation's forest resources;
3) the basic curriculum elements and level of instruction necessary to develop a core competence in the relevant knowledge base to produce suitably trained, socially aware, and technically proficient researchers and managers;
4) the means by which focused education and interdisciplinary systems thinking and communication skills can be developed and applied to forest and landscape problems; and

INTRODUCTION

5) the adequacy and capacity of available university programs to meet the needs of the near future.

A primary source of input for this study was a workshop that took place on July 15–16, 1999, at the National Academies building in Washington, D.C. In addition to the workshop and associated public comments and letters, the committee communicated with professionals in relevant forestry research and education organizations and consulted numerous other sources, including recent surveys and studies of trends in forestry education, to obtain relevant information for analysis.

THE CURRENT STUDY

This first chapter of the report describes the focus and boundaries of the committee's assessment, defines forestry research capacity, describes the institutional framework for forestry research, reviews the historic roots of forestry in the United States, addresses the continuing need for forestry research, and highlights future challenges involving forestry issues. Chapter 2 addresses the first charge of the committee and describes the essential knowledge base required by professionals who must address future needs, including education and research perspectives and priorities. Those priorities are classified into two broad areas of scientific need: foundation needs and emerging needs. Chapter 3 addresses the second charge of the committee and provides an overview of the current status of forestry-research capacity in terms of the resources that make up capacity: manpower, infrastructure, and financial investment. Chapter 4 addresses the third charge of the committee and looks at the status of forestry education, examines educational paradigms for graduate forestry education to produce the next generation of forestry researchers, and offers recommendations for enhancing the current status. Chapter 5 addresses the fifth charge of the committee synthesizes the material from the preceding chapters, assesses the status of our national research and education capacity with respect to priorities, and discusses various principles and approaches for meeting forestry-research needs. The fourth charge, which transcends several aspects of the overall assessment is addressed in Chapters 3 through 5.

In this report, the status of forestry-research capacity was assessed to determine whether desired social goals for the future could be reached. Research and monitoring provide the foundations required to improve management and protection and to achieve sustainable forest management. Heretofore, there has been a lack of adequate information on the magnitude of funds, personnel, and infrastructure that support forestry-research efforts. There has also been a lack of information on how current efforts were directed among disciplines; the magnitude of research capacity among federal, state, nongovernment, and private organizations; the breadth of forestry and forest-resources research and development; or the priorities for forestry research. Those issues are addressed in this report to the extent that data and resources allow.

Boundaries of the Assessment

There are major challenges associated with compiling a comprehensive information base needed to assess forestry research capacity. These challenges exist because information on forestry research and resources does not exist in a single place. Data reported—for example, on employment trends of forestry and natural resource researchers, are often reported for a certain sector and the methods or surveys used to collect and summarize the data across sectors often vary, making some comparisons impossible. Based on the limited amount of reliable information and the variation in data that were available to the committee for review and analysis, limits of the study had to be determined with regard to the current assessment of forestry research capacity.

In assessing the essential knowledge base for forestry issues, the boundaries were set as wide as possible and encompassed information received from public and private sources, universities with forestry and natural resource graduate and undergraduate programs, as well as industry, government and NGOs with forestry programs.

In assessing the capacity of research organizations to perform research, the boundaries were necessarily more narrow because information was more difficult to obtain. The review focused on the major forestry research agency of the federal government (the Forest Service) and other agencies in government for which there were data available on forestry research. The committee's assessment is focused on federal research that is uniquely federal and appropriately federal (see definitions provided in Chapter 5). Research performed in the forest industry and in academe was considered broadly, and focused on trends and apparent limitations.

In assessing curriculum elements, means for focused education and interdisciplinary systems, and adequacy of university programs to meet needs, the committee drew from the most comprehensive data sets available, including Food and Agricultural Education Information System (FAEIS) data and input from public and private institutions was gathered.

DEFINING FORESTRY-RESEARCH CAPACITY

To provide a perspective for this study, modern definitions of forestry and research capacity were used to guide the assessment. The Society of American Forester's (SAF) Dictionary of Forestry (Helms, 1998) defines forestry as:

> "The profession embracing the science, art, and practice of creating, managing, using, and conserving **forests** and associated resources for human benefit and in a sustainable manner to meet desired **goals, needs,** and values—note the broad field of forestry consists of those biologic, quantitative, managerial, and social sciences that are applied to **forest management** and **conservation**; it includes specialized fields such as agroforestry, urban forestry, industrial forestry, nonindustrial forestry, and **wilderness** and recreation forestry." (P. 72)

INTRODUCTION

Bengston and Gregersen (1988) provide a simple definition of research capacity:

"...as an institution's ability to develop and disseminate new technology."

In this report, forestry research is considered broadly. The classical forestry disciplines of biologic sciences, measurements, management, policy, and administration should clearly fall within the definition of forestry-research capacity. It would include forest insects and diseases, forest health, agroforestry, community forestry, spatial information, and a host of other disciplines applied to forestry. The published definitions were integrated in order to characterize forestry-research capacity for the committee's evaluation and assessment:

Forestry is the science, art, and practice of creating, managing, using, and conserving forests and associated resources in a sustainable manner, engaging broad and specialized scientific disciplines to meet desired goals, needs, and values.

Research capacity is the magnitude of the ability to develop, advance, and disseminate science and technology.

Because of the difficulty in obtaining historical data on forestry that transcend disciplines, the report focuses more on traditional tree and timber aspects of forestry and often does not address in detail the areas of fisheries, wildlife, water (quality and quantity), outdoor recreation, non-timber products, cultural resources, aesthetics, and forest social sciences in as great detail. However, these areas are recognized as important aspects of forestry and definitions of forestry have been broadening to include them; this concept is addressed in later chapters.

INSTITUTIONAL FRAMEWORK FOR
FORESTRY-RELATED RESEARCH

The institutional structure of forestry research in the United States consists of a number of different entities and this framework continues to broaden. The U.S. Department of Agriculture (USDA) Forest Service has been the major contributor to the nation's forestry-research portfolio, but many other federal departments and agencies are increasing support for research related to forests. These include the U.S. Environmental Protection Agency (EPA), the National Aeronautics and Space Administration (NASA), the U.S. Department of Energy (DOE), the U.S. Department of Defense (DOD), the U.S. Department of Interior (DOI), and the National Science Foundation (NSF). An example of the contribution that these organizations make to forestry research is the NSF-funded Long-Term Ecological Research (LTER) Network that has conducted forestry research for more than 20 years.

The nation's professional forestry schools, and more recently natural resources colleges, perform a large amount of the forestry research. Their faculties include people in forest sciences, such as biology, measurements, management, and policy, and those in emerging or related disciplines, such as ecology, environmental assessment, sociology, spatial information, and biotechnology. As a growing number of traditional forestry schools evolve into broader natural resources and environmental sciences schools, the research emphasis in them also shifts.

Forestry and natural-resources extension programs provide direct support for disseminating research findings to research users. Finally, forestry industry has more than one hundred scientific research personnel and also contributes to the country's capacity to conduct forest research.

Different research organizations and structures have different merits in generating knowledge. Forestry or any other research can be viewed as a continuum from a minimally controlled management system to a mission-oriented and tightly controlled approach (Roussopolous, 1999). Loosely controlled research organizations and approaches provide funds and a general charter to creative scientists, but little oversight and supervision. Louis Pasteur and classical scientific approaches might typify that approach. Such freedom and autonomy allow maximize creativity but guarantee neither success nor efficiency. Tightly controlled projects, such as the Manhattan Project, ensure that maximum effort and knowledge are brought to bear on solving a relatively well-defined problem. Such an approach is efficient but tends to restrict serendipitous discovery and pursuit of ancillary questions and findings.

The continuum posited by Roussopolous can be used to evaluate various forms of forestry-research organizations. Traditional mission-oriented research organizations, such as the USDA Forest Service, are most likely to achieve applied-research objectives where direction can be provided from within the agency and received from external research clients. Universities provide an environment where scientists may be more creative in selecting and performing research that can be funded. Forest industry has a directed research focus, which usually centers on topics that will improve the bottom line, including environmental topics related to ensuring sustainability and protection of asset values. Non-government organizations (NGOs) perform research that is moderately directed, whether toward management or specific concerns, such as environmental protection.

EARLY FORESTRY RESEARCH AND EDUCATION

Forestry research in the federal government of the United States has its roots in the early 1800s. In 1828, President John Quincy Adams germinated acorns in tubs around the White House with the notion that this method of cultivation of trees for timber could work in the field. Adams' live oak experiments were not carried to completion, but his efforts were among the first to stimulate interest in research on forests. Some 45 years later in, in 1873 Franklin B. Hough, a physician who would lead the U.S. government's first formal forestry-research efforts for almost a decade, made a presentation to the American Association for the Advancement of Science (AAAS) on the responsibilities of government in forestry research. In response to Hough's statements, AAAS petitioned Congress on the critical need to collect, compile, and distribute information on forest science and forested lands. Several years later, the first appropriations were made to the U. S. Department of Agriculture for forestry research.

In the same era, the first educational program in forestry was established in 1898 at Cornell University by a German forester, Bernard E. Fernow, who served as chief of the Division of Forestry in the Department of Agriculture for 12 years. Although short-lived, the program set the stage for several aspects of forestry education and research that remain today, including standardized curricula and a formal method of communicating research results.

THE IMPORTANCE OF MAINTAINING, PROTECTING, AND ENHANCING TODAY'S FORESTS FOR TOMORROW

In 1864, the cumulative impact of human actions on forests was summarized and applied to Europe and early America (Marsh, 1864). The scientific problem identified was that the nations of the world were expanding rapidly without apparent concern about effects on the land and its resources (Marsh, 1864). The problem persists today. Rapidly increasing world population and continuously decreasing forest area leave us with an enduring resource-management problem of getting greater benefits from less land. As population and per capita disposable income and consumption increase (World Resources Institute et al., 1998), demands for forest products increase. U. N. Food and Agriculture Organization (FAO, 1999) data indicate that 56 million hectares of forest was lost during the period from 1990 to 1995. That area—roughly equal to the total forest area of Western Europe—represents 1.6 percent of the world's forest area. The unprecedented increase in commodity and amenity demands from fewer forests creates significant pressure on the sustainability of the world's forest commodities and environmental benefits, and the concomitant community well-being.

Forests make up about 300 million hectares (747 million acres), or about 33 percent of the total land cover of 916 million hectares (2,263 million acres) in the United States (Smith et al., 2001). Estimates of the forest cover of the world vary. Williams (1994) summarized many studies of world forest area, noting that recent estimates had ranged from about 3.7 to 4.6 billion hectares. FAO (2001) estimated that forests cover 3.9 billion hectares, or about 26 percent of the earth's total land area of 13 billion ha.

Even current estimates of world and forest area depend on the definition of forests and the methods used to estimate the areas. For example, FAO (2001) indicates that the United States has only 226 million hectares of forest land, rather than the Smith et al. (2001) estimate of 300 million. The difference lies mostly in whether forests are defined as including only closed forests (FAO, 1999) or more-open forests with partial tree cover (Powell et al., 1993). Estimates of trends in forest area are probably less accurate. But most studies indicate that the forests of the world are being reduced in extent, at low to moderate rates. To further complicate issues, there has been a broadening of the definition of forestry. This greatly increases the scope of research that is needed to improve forest management.

As world forest area declines, world population continues to increase. The U.S. population in 1998 was about 274 million people, and demographers estimated that world population exceeded 6 billion in 1999. U.S. and world populations are projected to increase to 332 million and 8.0 billion people, respectively, by 2025 (World Resources Institute et al., 1998). World population increases have consistently slowed, but population has by no means stabilized except in some developed countries.

The preceding statistics frame the scenario for forests and all other natural resources on the earth. The total area of the earth covered by forests is declining while the number of people demanding products, goods, and services (see Table 1) from forests is increasing, as is their per capita consumption of commodities and amenities. Developed countries use more forest products for housing, printing and writing, packaging, personal products, and business. Similarly, the aging and more affluent and mobile populations, at least in the Northern Hemisphere, place more demands for recreation and environmental benefits on forests throughout the world. Thus, forests throughout the United States and the world are under increasing pressure to provide sustained and, in fact, increased outputs. Increased demands from forests include those for wood products, high-quality water, and intensive and passive recreation. Forests also provide sites for urban expansion and recreational homes, wildlife habitat and biodiversity, carbon storage and oxygen production, and a variety of indirect benefits.

Natural forests have diverse flora and fauna and provide scenic beauty, carbon storage and oxygen production, forest products, and other benefits. Planted forests and trees provide industrial roundwood, fuel for homes, and urban amenities. Forest reserves protect natural values, biodiversity, and ecologic integrity (see Table 1-1). However, forests—even reserves—are threatened by encroachment of humans and nonnative flora and fauna. Ecologic restoration and management might make it possible to maintain or restore natural conditions.

Table 1-1. Definitions of Terms Commonly Used to Describe What Forests Provide.

Term	Definition	Example
Products	Things, substances, articles produced by a process; output of goods and services resulting from the input of resources or factors of production used to produce them.	Paper products
Goods	Things, articles, objects worth attaining; movable properties; merchandise; wares; services of valve. An economic good is defined as any physical object, natural or man-made, or service rendered, which could command a price in a market.	Timber
Services	Provision of assistance; act of serving; work done to meet some needs; intangible, non-transferable economic goods, as distinct from physical commodities.	Clean air, clean water
Values	Enjoying the forest for the forest	Personal appreciation, importance placed on the forest
Benefits	Advantage; favorable effect; output; profit. Includes products and favorable influences.	Income
Output	Similar to products but is often compared to inputs.	Production measure

Source: Adapted from http://www.fao.org/docrep/V7540e/V7540e28.htm

FUTURE CHALLENGES

The increasing demands for forest goods and services show the importance of the role of forestry education and research to support sustainable management and conservation. The "Brundtland report" (World Commission on Environment and Development, 1987) defines sustainable development as "development that meets the needs of the present without compromising the ability of future generations to meet their own needs". This principle has been widely applied to efforts to use and protect the world's forests. How are we to reach the laudable, but difficult, goal of sustainable development? What are the precepts of sustainability? Can we not only sustain our resources, but improve their condition? Elements of philosophy, management, education, research, and practice are all involved in achieving sustainable forest management.

The world must strive to realize the three tenets of sustainable development: economic development, environmental protection, and community welfare. The public is demanding better environmental performance from private firms and from public managers, as well as high product quality at reasonable costs. Forest-management objectives vary among ownerships, but all are seeking stewardship, sustainable forest management, and cost minimization, if not economic profitability. Private and public managers in the future must merge economic and environmental goals in producing goods and services and in reducing resource consumption and pollution per unit of output (World Resources Institute et al., 1998). Forest-management intensity will differ among landowner classes as well, ranging from intensive to extensive management practices and commodity production to natural-forest protection. Research in and implementation of methods to improve forest use also offer means to conserve natural resources and protect the environment.

Manufacturing and leisure uses of resources should not degrade their quality and businesses should reduce the use of material and energy and the production of waste. Research programs must move towards discoveries of highly productive and environmentally benign processes, and professionals of the future must receive the best science education and technology transfer to benefit people throughout the United States and the world.

FORESTRY EDUCATION AND RESEARCH

At its core, achieving sustainable forest management will require greatly increasing our research capacity. Increasing capacity begins with preparing the forestry professionals of the future with a solid educational foundation to assess the status of our forests, finding means to enhance their values for commodity and noncommodity outputs, and implementing improved management or protection of the relevant forest areas. Monitoring, research, and application of new technology must be the bases for sustaining, enhancing, and restoring forests and their innumerable values. In achieving sustainability, many people call for expanding the traditional forestry view of "sustainable yield" practices to a concept of "sustaining ecosystems" (Noble and Dirzon, 1997).

Better monitoring, research, and application are paramount in enhancing forest productivity and protection and they constitute the challenge: How do we enhance education and research efforts and applications—from basic to applied research and from simple to elegant technology transfer—and improve forest protection and management? How do we measure and monitor progress? Is our capacity in forestry education and research up to the task? In fact, is our capacity in forestry education and research even maintaining historical levels, let alone providing the ability to increase our knowledge about forests?

Those questions are integral to this study by the National Research Council. As the request for such a study suggests, the nation's research capacity in forestry—even to maintain existing programs and knowledge—is in question. Whether we have adequate forestry-research capacity to increase productivity or protection is moot. The implied fears might be unfounded.

For decades, the National Research Council has shed light on these difficult topics pertaining to forestry research (National Research Council, 1926, 1927a,b, 1928, 1947, 1990). However, there has not been a comprehensive assessment of U.S. forestry-research capacity since *Mandate for Change* was released more than 10 years ago (National Research Council, 1990). Trends in forestry education have recently been assessed (Pinchot Institute for Conservation, 2000), but the integrated impacts of education and research on the nation's capacity in forestry research have not been followed closely or appraised lately. A fresh look is warranted, inasmuch as traditional forestry education and research entities continue to be called on to meet vast challenges and many organizations not traditionally considered to be dedicated to forestry education and research now contribute largely to such activities.

2

The Essential Knowledge Base for Forestry Issues

Public and private forest managers, related professionals, citizen conservationists, and officials responsible for forest policy all perform optimally when their actions are based in part on knowledge derived from research. Many forestry-related decisions and actions required of individuals and institutions have become much more complex as new management paradigms, such as ecosystem management and ecologic sustainability, have been developed to address growing societal concerns.

Ensuring an adequate and appropriate knowledge base to address current and future needs in forestry research requires an integrated, coordinated approach to education and research. A 1964 report by the National Research Council stated that "education can be of highest quality only if it is conducted as part of the research process itself" (National Research Council, 1964). There is general agreement in the forestry profession that a graduate degree, usually a doctorate, is necessary if one wishes to conduct specialized forestry research (Pinchot Institute for Conservation, 2000). By educating students in the context of research, the American system of graduate education has set the world standard for preparing scientists for research careers in academe, government, private industry, and other organizations. In forestry, as in most sciences, education and research are often inextricably linked. Emerging knowledge needs will be met by education and research that address complex social, political, and technical issues. Forestry education and research today should be interdisciplinary and visionary in anticipating societal needs. Success depends on essential foundation knowledge that has been the backbone of forest management for many years. Foundation knowledge is the key to meeting society's ever-growing demand for forest-based products.

Over the last decade, a number of studies have identified and recommended new directions—actually expanded directions more than replacement directions— for forestry education and research (American Forest Congress, 1996; Ginger et al., 1999; Ellefson and Ek, 1996; Committee of Scientists, 1999). Much of the knowledge base needed to address forest-related issues in the early decades of the 21st century has been identified and enumerated in these studies. In addition, advice solicited from experts for the present report revealed that the following topics have the highest priority.

Foundation forestry education and research:

- Biology, ecology, and silviculture
- Forest genetics
- Forest management, economics, and policy
- Wood and materials science

Emerging forestry education and research:

- Human and natural resource interactions
- Ecosystem function, health, and management
- Forest systems on various scales of space and time
- Forest monitoring, analysis, and adaptive management
- Forest biotechnology

The division into priorities among foundation and emerging forestry education and research reinforces the needs for traditional education and research functions, but in a new context, and for extending our knowledge to relatively new disciplines, which are rapidly becoming more important. Foundation programs are required as the base on which new research will be built. Furthermore, the basic programs and applied forestry sciences are becoming far better grounded and supported by advances in the emerging fields. It is essential that decision-makers in both the public and the private sectors understand and support these dual forestry education and research needs.

Basic programs and basic science (especially foundation research) have made exceptional advances in rigor, depth of understanding, and potential for enhanced sustainable forest management in the last decade. Basic biotechnology, genomic science, tree breeding, and forest physiology promise to allow us to triple or quintuple rates of forest growth per unit area and to impart disease and insect resistance on intensively managed lands. Foundation research also can allow us to select trees that have desirable wood properties and to manipulate their cell and wood structure (genetics) and environment to produce more uniform wood with targeted wood properties for industrial applications. Foundation genetics education and research allow us to examine questions of genetic diversity at the cell, stand, or landscape level. Sustainable forest management promises to focus efforts on extending traditional sustained-yield forestry to multiple scales, periods, goods and services, and forest interest groups; this will increase our ability to provide economic development, ensure environmental protection, and improve community welfare.

Emerging forestry education and research have become more important over the last decade. To some extent, these disciplines constitute not new fields, but rather new and more integrative ways of viewing foundation disciplines and their applications to forest-resource management. An extensive review and summary of scientific disciplines important to forestry and renewable resources published in 1983 remains relevant today (University of Idaho, 1983). Human and natural resource interactions have always been important, but are stressed more as the number of people increases and the forest area decreases. Ecosystem function and management were initiated as a public-land management paradigm a decade ago and have become widely adopted on public lands since. Forest systems at the landscape, national, and international levels—for present and future generations—have become more important. As sustainable forest management becomes a widespread forestry paradigm, measuring and monitoring the status of forests and progress in enhancing forest condition become more important.

KNOWLEDGE BASE REQUIRED

The foundation and emerging priority education and research areas integrate the literature, comments, and workshop input that the study committee received regarding the knowledge base required to ensure adequate future national capacity in forestry research. In fact, the information gathered on the knowledge base required essentially defined a broad set of forestry issues that merit further study. In this chapter, some of the critical education and research needs are summarized that were identified throughout the study process; curriculum needs and models for education are addressed in depth in Chapter 4. Choosing and classifying high-priority education and research needs is a somewhat presumptuous and daunting task, given the wealth of published literature, public discussions and forums, and interest groups that have a stake in such questions. For simplicity, some of the principal education and research priorities issued by selected processes and interest group are reviewed, summarized, and then assessed in accordance with the categories provided above. A similar process was used with regard to comments provided to the committee during its July 15–16, 1999, workshop (see Appendix A for the workshop agenda and Appendix B for the breakout-group discussion topics).

Necessary knowledge will be acquired through education that provides the intellectual resources for future forestry research based on well-defined priorities. The National Research Council (1990) listed five research priorities—biology of forest organisms, ecosystem function and management, human-forest interactions, wood as a raw material, and international trade, competition, and cooperation. The 1996 American Forest Congress identified research priorities for each major region of the country. The Northeast Region used the 1990 National Research Council report priorities as a template for its 1996 congress. The other four regions developed their own research priorities. The forest industry developed research priorities through their participation in the Department of Energy Agenda 2020 competitive research process. The Forest Service Forest Experiment Stations have developed strategic plans that outline research priorities for their regions. Table 2-1 summarizes research and education priorities according to various sources. They suggest a wide variety of research and supporting education needs for the nation, and they are discussed below in the context of foundation and emerging education and research priorities.

Foundation Education and Research Priorities

Each of the organizations that identified research priorities had a mix of activities in each of the four categories of foundation, forestry, education, and research presented at the beginning of this chapter. On the basis of the education and research priorities listed in Table 2-1, one could consolidate subjects in the four categories:

- *Biology, ecology, and silviculture.* Biology of forest organisms, sustainable forest productivity, long-term soil productivity, basic physiology of forest trees biochemistry, and proactive pest management.
- *Forest genetics.* Genetics, identification of major traits, tree breeding, genetics by environment interactions..
- *Forest management, economics, and policy.* Sustainable forestry and land stewardship, impact of regulations on forestry, livestock grazing, fire, management policies for public lands, socioeconomic and political factors and effectiveness, urban-rural issues, improvement in productivity and harvesting technology, decision-support and management models and international trade, competition, and cooperation.
- *Wood and materials science.* Wood as raw material, forest use, and wood and fiber production.

Emerging Education and Research Priorities

On the basis of the priorities listed in Table 2-1, one could consolidate the following subjects in the five categories of emerging forestry education and research presented at the beginning of this chapter:

- *Human and natural resource interactions.* Human-forest interactions; social-science methods; enrichment of recreation user experiences; communication of results to users; infrastructure development; economic, regulatory, and demographic factors; small landowners' production of high-value and special products; economics of nontimber resources; and wilderness, recreation, tourism, and aesthetics.
- *Ecosystem function, health, and management.* Ecosystem structure, function, process, and management; water quality and forested wetlands and protection; enhancement of health and productivity; rehabilitation and recovery efforts; wildlife habitat in managed forests; and biodiversity, ecosystem management, and adaptive management.
- *Forest systems on various scales of space and time.* Water, watersheds, and riparian zones; cumulative effects; and climate-change impacts.

Table 2-1. Comparative Forestry-Science Education and Research Priorities According to Selected Sources.

Education				Research		
Pinchot Institute for Conservation (2000)	NRC (1990) & American Forest Congress: Northeast (1996)	American Forest Congress: South (1996)	American Forest Congress: Lake States (1996)	American Forest Congress: Pacific Northwest (1996)	American Forest Congress: Pacific Southwest (1996)	Forest Industry Agenda 2020 (1996)
• Silviculture systems	• Biology of forest organisms	• Sustainable forest productivity	• Monitoring resources	• Economic, regulatory, and demographic factors	• Demographic characteristics	• biotechnology to locate important quantitative genes
• Forest ecology	• Ecosystem function and management	• Water quality and forested wetlands	• Enhancing health and productivity	• Small loss production of high-value products	• Socioeconomic and political factors	• biotechnology tools for genetic transformation in trees
• Forest inventory and biometry	• Human-forest interactions	• Decision support and management models	• Improving productivity and harvesting technology	• Rehabilitation and recovery efforts	• Urban-rural issues	• tissue-culture technology for trees
• Species identification	• Wood as raw material	• Forest inventory, analysis, and growth	• Protection of water resources	• Wildlife habitat in managed forests	• Management policies for public lands	• physiology research to accelerate plant growth
• Forest soils	• International trade, competition, and cooperation	• Ecosystem structure, function, and process	• Enrichment of recreation user experiences	• Climate-change impacts	• Sustainability and land stewardship	• physiology to support genetic engineering
• Wildlife biology		• Social-science methods	• Improved raw-material use	• Biodiversity, ecologic management, and adaptive management	• Forest health and biodiversity	• physiology and forest and ecosystem management
• Communication		• Forest use	• Enhancing resource management policy effectiveness	• Long-term soil productivity	• Fire	• sustainable soil productivity
• Ethics			• Communication of results to users	• Watershed cumulative effects	• Wood and fiber production	• intensive management effects and site productivity
• Leadership			• Infrastructure development	• Systems to integrate resource information	• Wildlife habitat and endangered species	• soil limits on site productivity
• Collaborative problem-solving				• Basic physiology of forest trees	• Livestock grazing	• treatments to enhance soil productivity
				• Impact of regulations on industry	• Recreation, tourism, and aesthetics	• improvement in forest-inventory program
				• Sustainable forestry	• Water, watersheds, riparian	• monitoring of forest health and productivity
				• Proactive pest management	• Special products	
				• Economics / nontimber resources		

- *Forest monitoring, analysis, and adaptive management.* Inventory methods and resource analysis, monitoring of resources, remote sensing and geographic information systems (GIS), and systems to integrate resource information.
- *Forest biotechnology.* Biotechnology, location of important quantitative genes, tools for genetic transformation, and tissue culture.

Stewardship and Sustainability of Public Lands

A recent analysis of the management needs for national forests provided a way to examine research needs for public lands. The Committee of Scientists (COS) report, *Sustaining the People's Lands* (1999) calls for the stewardship of the national forests and grasslands to be guided by principles of sustainability and the recognition that these are the people's lands. To achieve ecologic sustainability, the report articulates a challenging list of principles, including acknowledgment of the dynamic nature of ecologic systems and the identification of human uses that contribute to long-term sustainability. All the principles have profound implications for research.

For some of the COS principles, a considerable body of work has been done to understand and practice ecosystem management. An example is "involve the scientific community in developing strategies for maintaining ecologic, economic, and social sustainability." Other principles involve new concepts for research, such as "recognizing that planning and management of public lands proceeds under legitimate but often divergent, interests." This seems to speak to the heart of what must be understood if the American public is to respect and support forest management as an essential part of society. *Sustaining the People's Lands* offers a blueprint for identifying many of the complex (multidisciplinary) kinds of research needed to provide the knowledge base for 21st century public-land forestry management. Similarly, the host of reports that have been written on ecosystem management identifies numerous complex subjects that need research.

Sustaining the People's Lands overlaps substantially with *Forestry Research: A Mandate for Change* (National Research Council, 1990) in recommended actions. The need for knowledge in human-forest interaction (these are the people's lands) and ecosystem function and management (ecologic sustainability) presents priorities that have met the dual tests of time and separate review.

Sustainable-Management Criteria and Indicators

Perhaps the most integrative list of research topics for the future might be the sustainable forest management (SFM) criteria and indicators that have been promulgated for most of the forests in the world and agreed on through various international treaties. At the June 1992 U.N. Conference on Environment and Development in Rio de Janeiro (the "Earth Summit"), 144 countries developed and adopted a nonbinding "Statement of Forest Principles" that recognized the importance of SFM for all types of forests. In 1993, a U.N. committee meeting on sustainable development of temperate and boreal forests was held in Montreal. That meeting included representatives of nine nations who formed the Working Group on Criteria and Indicators for the Conservation and Sustainable Management of Temperate and Boreal Forests. The "Montreal Process" countries met in Santiago, Chile, in February 1995 to endorse their commitment to the Montreal process. The "Santiago declaration" accepted a comprehensive set of seven criteria and 67 indicators for the conservation and sustainable management of temperate and boreal forests. Similar criteria and indicators for measuring and assessing SFM were developed through the "Helsinki process" in Europe. Earlier efforts by the International Tropical Timber Organization also were designed to enhance SFM (National Association of State Foresters, 1997).

There are now 130 countries engaged in activities related to criteria and indicators. As of 1997, the 12 Montreal process countries were Argentina, Australia, Canada, Chile, China, Japan, the Republic of Korea, Mexico, New Zealand, the Russian Federation, the United States, and Uruguay. Those countries are on five continents and contain 90 percent of the world's temperate and boreal forests and 60 percent of all forest on the globe. The Montreal process developed broad *criteria* that were intended to represent a large-scale reflection of public values; *indicators* were then developed to provide a means of measuring forest conditions and tracking changes (National Association of State Foresters, 1997).

The SFM criteria and indicators might be considered the raison d'être for forest research, monitoring, and adaptive management. Table 2-2 summarizes the seven criteria and 67 indicators promulgated under the Montreal process. Applying them on public and private lands in the United States and in other countries clearly poses a huge challenge for forestry in the future; education and research will be crucial. The breadth of the criteria encompasses virtually all forest practices, from biologic to social questions. They also could fall within our foundation and emerging priorities, or vice versa. The widespread international agreement on the criteria suggests that they will become the paradigm governing forest management, education, and research.

Note that five broad SFM criteria generally address biologic, physical, or natural standards for forest management and that only two address economic, social, or institutional issues. However, only 28 of the indicators address biophysical standards, and 39 social and institutional standards. Measuring many of the qualities—biophysical or social—is extremely difficult. The U. S. Department of Agriculture Forest Service has estimated that fewer than half of the indicators are directly measurable with current monitoring approaches and technologies, and current public funding levels.

Monitoring and research comprise the last set of eight indicators under the legal and social criteria. These indicators are related to how well we measure and monitor the preceding quantities and to whether we are making progress toward SFM. The United States and all other member countries of the Montreal process are mandated to do a better job of measuring and monitoring the first 59 indicators and explicitly charged with improving monitoring and research in the last eight indicators. These countries are committed to preparing the first full report on SFM and meeting the criteria and indicators standards in 2003.

Table 2-2. Criteria for the Conservation and Sustainable Management of Temperate and Boreal Forests Developed from the Montreal Process.

	Criteria	Description
1	Conservation of biologic diversity	*Ecosystem diversity:* 1) Extent of area by forest type relative to total forest area 2) Extent of area by forest type and by age class or successional stage 3) Extent of area by forest type in protected-area categories as defined by IUCN or other classification systems 4) Extent of area by forest type in protected areas defined by age class or successional stage 5) Fragmentation of forest types *Species diversity:* 6) Number of forest-dependent species. 7) Status (rare, threatened, endangered, or extinct) of forest-dependent species at risk of not maintaining viable breeding populations, as determined by legislation or scientific assessment *Genetic diversity:* 8) Number of forest-dependent species that occupy a small portion of their former range 9) Population levels of representative species from diverse habitats monitored across their ranges
2	Maintenance of productive capacity of forest ecosystems	10) Area of forest land and net area of forest land available for timber production 11) Total growing stock of both merchantable and nonmerchantable tree species on forest land available for timber production 12) Area and growing stock of plantations of native and exotic species 13) Annual removal of wood products compared with the volume determined to be sustainable 14) Annual removal of nontimber forest products (such as fur bearers, berries, mushrooms, game) compared with the level determined to be sustainable
3	Maintenance of forest ecosystem health and vitality	15) Area and percentage of forest affected by processes or agents beyond the range of historic variation (for example, insects, disease, competition from exotic species, fire, storm, land clearance, permanent flooding, salinization, and domestic animals) 16) Area and percentage of forest land subjected to levels of specific air pollutants (for example, sulfates, nitrate, and ozone) or ultraviolet B that can cause adverse effects on the forest ecosystem 17) Area and percentage of forest land with diminished biologic components indicative of changes in fundamental ecologic processes (such as soil, nutrient cycling, seed dispersion, and pollination) or ecologic continuity
4	Conservation and maintenance of soil and water resources	18) Area and percentage of forest land with significant soil erosion 19) Area and percentage of forest land managed primarily for protective functions (such as, watersheds, flood protection, avalanche protection, and riparian zones) 20) Percentage of stream kilometers in forested catchments in which stream

	Criteria	Description
		flow and timing have significantly deviated from the historic range of variation
		21) Area and percentage of forest land with significantly diminished soil organic matter or changes in other soil chemical properties
		22) Area and percentage of forest land with significant compaction or change in soil physical properties resulting from human activities
		23) Percentage of water bodies in forest areas (such as stream kilometers and lake hectares) with significant variance of biologic diversity from the historic range of variability
		24) Percentage of water bodies in forest areas (such as stream kilometers and lake hectares) with significant variation from the historic range of variability in pH, dissolved oxygen, levels of chemicals (electric conductivity), sedimentation, or temperature change
		25) Area and percentage of forest land experiencing an accumulation of persistent toxic substances
5	Maintenance of forest contribution to global carbon cycles	26) Total forest ecosystem biomass and carbon pool and if appropriate, by forest type, age class, and successional stages
		27) Contribution of forest ecosystems to the total global carbon budget, including absorption and release of carbon
		28) Contribution of forest products to the global carbon budget
6	Maintenance and enhancement of long-term multiple socio-economic benefits to meet the needs of societies	*Production and consumption:*
		29) Value and volume of wood and wood-products production, including value added through downstream processing
		30) Value and quantities of production of nonwood forest products
		31) Supply and consumption of wood and wood products, including consumption per capita
		32) Value of wood and nonwood-products production as percentage of gross domestic product
		33) Degree of recycling of forest products
		34) Supply and consumption or use of nonwood products
		Recreation and tourism:
		35) Area and percentage of forest land managed for general recreation and tourism in relation to the total area of forest land
		36) Number and type of facilities available for general recreation and tourism in relation to population and forest area
		37) Number of visitor-days attributed to recreation and tourism in relation to population and forest area
		Investment in the forest sector:
		38) Value of investment, including investment in forest growing, forest health and management, planted forests, wood-processing, recreation, and tourism
		39) Level of expenditure on research and development and or education
		40) Extension and use of new and improved technology
		41) Rates of return on investment
		Cultural, social, and spiritual needs and values:
		42) Area and percentage of forest land managed in relation to the total area of forest land to protect the range of cultural, social, and spiritual needs and values

Criteria	Description
	43) Non-consumptive-use forest values

Employment and community needs:
44) Direct and indirect employment in forest-sector and the forest sector employment as a proportion of total employment
45) Average wage rates and injury rates in major employment categories in the forest sector
46) Viability and adaptability to changing economic conditions of forest-dependent communities, including indigenous communities
47) Area and percentage of forest land used for subsistence purposes

| 7 | Legal, institutional, and economic framework for forest conservation and sustainable management | *Extent to which the legal framework (laws, regulations, and guidelines) supports the conservation and sustainable management of forests, including the extent to which it*
48) Clarifies property rights, provides for appropriate land-tenure arrangements, recognizes customary and traditional rights of indigenous people, and provides means of resolving property disputes by due process
49) Provides for periodic forest-related planning, assessment, and policy review that recognize the range of forest values, including coordination with relevant sectors
50) Provides opportunities for public participation in public policy-making and decision-making related to forests and public access to information
51) Encourages best-practice codes for forest management
52) Provides for the management of forests to conserve special environmental, cultural, social, and scientific values

Extent to which the institutional framework supports the conservation and sustainable management of forests, including the capacity to
53) Provide for public involvement and public education, awareness, and extension programs and make forest-related information available
54) Undertake and implement periodic forest-related planning, assessment, and policy review, including cross-sectoral planning and coordination
55) Develop and maintain human-resource skills across relevant disciplines
56) Develop and maintain efficient physical infrastructure to facilitate the supply of forest products and services and to support forest management
57) Enforce laws, regulations, and guidelines.

Extent to which the economic framework (economic policies and measures) supports the conservation and sustainable management of forests through
58) Investment and taxation policies and a regulatory environment that recognize the long-term nature of investments and permit the flow of capital in and out of the forest sector in response to market signals, nonmarket economic valuations, and public-policy decisions to meet long-term demands for forest products and services
59) Nondiscriminatory trade policies for forest products

Capacity to measure and monitor changes in the conservation and sustainable management of forests, including
60) Availability and extent of up-to-date data, statistics, and other information important for measuring or describing indicators associated with criteria 1- |

Criteria	Description
	7
	61) Scope, frequency, and statistic reliability of forest inventories, assessments, monitoring, and other relevant information
	62) Compatibility with other countries in measuring, monitoring, and reporting on indicators
	Capacity to conduct and apply research and development aimed at improving forest management and delivery of forest goods and services, including
	63) Development of scientific understanding of forest ecosystem characteristics and functions
	64) Development of methods of measuring and integrating environmental and social costs and benefits into markets and public policies and of reflecting forest-related resource depletion or replenishment in national accounting systems
	65) New technologies and the capacity to assess the socioeconomic consequences associated with the introduction of new technologies
	66) Enhancement of ability to predict impacts of human intervention on forests
	67) Ability to predict impacts of possible climate change on forests

A Roundtable on Sustainable Forests (http://www.sustainableforests.net/) has been formed to develop means of measuring and monitoring progress in achieving the criteria and indicators standards and of preparing the report in 2003; several meetings helped define terms and to develop measurement protocols. The international agreements and national implementation committees promise to make SFM criteria and indicators central in forestry research, monitoring, and adaptive management.

Forest Certification

Forest certification is a rapidly developing new means to enhance forest management and protection, and potentially generate adequate financial returns from working forests to ensure that they are retained. Various certification approaches exist, but the two dominant systems in the United States are the Sustainable Forestry Initiative (SFI), administered by the American Forest and Paper Association (AFPA), and the Forest Stewardship Council (FSC) approaches. The broad principles that these two certification approaches mandate also provide a telling framework for the various important research foci in the future. These or similar approaches will be applied to much of the managed forests in the United States.

There has been a rapid increase in the areas of forests that have been certified since those systems began in 1993. As of 2000, about 94 million hectares of forest were certified in the world by one of the certification systems (FAO, 2001, Meridian Institute, 2001). In the United States, by 2001 the Forest Stewardship council had granted 64 certified forest certificates, covering 3.3 million hectares, as well as issued 391 chain-of-custody certificates. FSC had certified 22 million hectares in the world. By 2001, the Sustainable Forestry Initiative had 132 company participants and 52 licensees, owning 14 million hectares in the United States. It had granted 16 third-party certification actions, covering 8 million hectares in the U.S. SFI also approved certifications in Canada, which included 14 million hectares more (Meridian Institute, 2001).

It is not possible to list all the criteria and a standard, since they are at least a dozen pages each. SmartWood has developed generic Principles with many Criteria to measure the FSC Principles listed in Forest Stewardship Council (2001).

1) *Compliance with Laws and FSC Principles.* Forest management shall respect all applicable laws of the country in which they occur, and international treaties and agreements to which the country is a signatory, and comply with all FSC Principles and Criteria.

2) *Tenure and Use Rights and Responsibilities.* Long-term tenure and use rights to the land and forest resources shall be clearly defined, documented, and legally established.

3) *Indigenous People's Rights.* The legal and customary rights of indigenous peoples to own, use and manage their lands, territories, and resources shall be recognized and respected.

4) *Community Relations and Worker's Rights.* Forest management operations shall maintain or enhance the long-term social and economic well being of forest workers and local communities.

5) *Benefits from the Forest.* Forest management operations shall encourage the efficient use of the forest's multiple products and services to ensure economic viability and a wide range of environmental and social benefits.

6) *Environmental Impact.* Forest management shall conserve biologic diversity and its associate values, water resources, soils, and unique and fragile ecosystems and landscapes, and, by so doing, maintain the ecologic functions and integrity of the forest.

7) *Management Plan.* A management plan–appropriate to the scale and intensity of the operations–shall be written, implemented, and kept up to date. The long-term objectives of management, and the means of achieving them, shall be clearly stated.

8) *Monitoring and Assessment.* Monitoring shall be conducted–appropriate to the scale and intensity of forest management–to assess the condition of the forest, yields of forest products, chain of custody, management activities and their social and environmental impacts.

9) *Maintenance of High Conservation Value Forests.* Management activities in high

conservation value forests shall maintain or enhance the attributes, which define such forests. Decisions regarding high conservation value forests shall always be considered in the context of a precautionary approach.

10) *Plantations.* Plantations shall be planned and managed in accordance with Principles and Criteria 1 - 9, and Principle 10 and its criteria. While plantations can provide an array of social and economic benefits, and can contribute to satisfying the world's needs for forest products, they should complement the management of, reduce pressures on, and promote the restoration and conservation of natural forests.

In total, there are 10 FSC Principles and 56 Criteria. In addition, there are may be more than 200 "bulleted" standards that certifying organizations use to provide details as to how the Principles and Criteria should be measured. This provides a detailed and rigorous set of standards to measure forest management. Principle #10 (Plantations) presents some of the more challenging components for high intensity forestry, requiring clear justification, protection of natural forests, species diversity, and long-term site protection. Plantations established in areas converted from natural forests after November 1994 normally shall not qualify for certification unless the current owner was not responsible for the conversion.

The SFI criteria were last revised in 2001 (American Forest & Paper Association, 2001), and require a set of core indicators for all organizations for a program participant to successfully complete 3rd party verification. The SFI standards also allow organizations to select optional indicators that they consider appropriate for their management systems and conditions. In total, there are about 100 core SFI indicators; there can be dozens to more than a hundred optional criteria at an organization's election. The paraphrased core objectives and standards are:

1) Broaden the practice of sustainable forestry, including having a written policy or program; providing funding for forest research; provide recreation and education opportunities; and ensure that long-term harvest levels are sustainable.

2) Ensure long-term forest productivity and reforestation, through reforestation by natural or planted methods within two years; promote state-level reporting of the overall rates of reforestation success and afforestation; use chemicals prudently and follow BMPs; implement management practices to protect and maintain soil productivity; protect forests from damaging insects, diseases, or fires; and use genetically improved material with sound scientific methods.

3) Protect water quality by using BMPs developed under EPA approved state water quality programs and meet or exceed all state water quality laws; develop, implement, and document riparian protection measures; provide funding for water quality research; and require BMP training for company employees in woodlands and procurement, and encourage training for forest management and harvesting contractors.

4) Manage the quality and distribution of wildlife habitat, and contribute to the conservation of biologic diversity by having programs and plans to promote habitat diversity at the stand and landscape level; fund research; apply research and technology and practical experience in wildlife and biodiversity management.

5) Manage the visual impact of harvesting and other forest operations, through planing and design; managing the size of clearcuts, with an average size not to exceed 120 acres; adopting a 3-year (5') green-up requirement before adjacent areas ay be clearcut; and vary harvest units to promote diversity.

6) Manage lands of ecologic, historic, and geologic significance carefully.

7) Promote the efficient use of forest resources, by minimizing waste and ensuring efficient utilization in the woods.

8) Cooperate with forest landowners, wood producers, and consulting foresters, by encouraging use of BMPs and providing environmental and economic information about BMPs; working closely with state logging and/or forestry associations and agencies.

9) Publicly report progress in fulfilling their objective to sustainable forestry.

10) Provide opportunities for the public and forestry community to participate in the commitment to sustainable forestry.

11) Promote continual improvement in the practice of sustainable forestry and monitor, measure, and report performance in achieving the commitment to sustainable forestry.

Forest-Industry Priorities

The AFPA Agenda 2020 process defined similar sustainable-forestry research priorities for the forest-products industry (American Forest and Paper Association, 1999). These included the broad categories of biotechnology, basic physiology of forest productivity, sustainable forest productivity, and remote sensing technologies to improve forest inventory and stand management. The Agenda 2020 process also included wood-science manufacturing research issues such as environmental performance, energy performance, recycling, and sensors and controls. Forest-industry firms have also released research priorities via various regional committees. Most of these priorities focus on research that supports intensive production of wood while sustaining the productivity of the land; research should develop baseline data on forest inventory, timber supply, environmental forestry, and water issues (i.e., AFPA North Central Forest Resources Research Committee, 1996; Southern Industrial Forestry Research Council,

1996). AFPA believes that the nation's ability to address research priorities adequately will depend on the education, production, procurement, and support of scientists in the public and private sectors.

New Forestry-Research Challenges

In addition to the move toward sustainable indicators and timber certification that require a broader and more interdisciplinary approach to forestry, there is an exciting and demanding 'new world' of forestry research, which requires expanding the knowledge base of forest scientists and managers. These new challenges suggest that scientists be prepared to move in broader, more interdisciplinary directions and examples include:

a) The science implications of policies of sustainability as referred to in the Brundtland report, and long-term forest productivity.

b) The trend for policy and management decisions to be driven by science (e.g., FEMAT, the Interior Columbia River Basin Study, NW Forest Plan, Congressional Hearings on Forest Health, and Southern Forest Resource Assessment).

c) Potential global climate change, carbon sequestration, and carbon credits.

d) Effects of potentially eliminating harvesting of timber on public lands on issues of supply and demand (regional, national, and global), forest health, and recycling forest products.

e) Biotechnology and yield improvement.

f) New management/science issues including: forest health, wildfire and fuels reduction, integrated pest management, watershed assessment, adaptive management, habitat for threatened and endangered species, multiple-species wildlife habitat, effects of forest fragmentation, watersheds and riparian issues, medicinal/food plants and other on-timber forest products, fire surrogates, variable retention silviculture, urban forestry, agroforestry, community well-being, structural dynamics, spatial/temporal issues, and inventory and monitoring.

These issues are creating a new paradigm for forestry research and are driving, in part, the need to assess our current forestry research capacity. The critical nature of the new world of forestry places urgency on assessing and ensuring a strong national research capacity for the future.

WORKSHOP INPUT ON AN
ESSENTIAL KNOWLEDGE BASE

Before, during, and after the 1999 National Research Council Workshop on Forestry Research Capacity, the committee received presentations and written comments (see Box 2-1) from representatives of universities (22), government agencies (8), trade organizations (4), non-government organizations (3), and private industry and others (3).

Box 2-1 Excerpts from Input Received on Education and Research Needs to Form an Essential Knowledge Base

"We may lose sight of the required basic forestry skills, . . . the most important set of skills. . . . Some erosion of these basic skills has already occurred." —Sam E. Curl, dean and director, Division of Agricultural Science and Natural Resources, Oklahoma State University

"Silviculture and forest ecology are just as important as studying social science topics like economics and policy."— Bobby D. Moser, dean, College of Food, Agricultural and Environmental Sciences, and Gary W. Mullins, director, School of Natural Resources, The Ohio State University

"Knowledge base . . . requires an extraordinary investment in research that addresses the ecosystem in which the forest exists."—James J. Zuiches, dean, College of Agriculture and Home Economics, Washington State University

"The following knowledge base components are important and necessary.
1) Forest resource (timber and non-timber) inventory, including assessment and prediction of forest resource inventory, dynamics, response to management practices and other disturbances, and interactions between forest resource and atmospheric changes.
2) Public needs/demand and desires for forest resources (timber and non-timber products).
3) The diversity of forest landowners—public, industrial, and non-industrial private (including socially disadvantaged) landowners.
4) Scientific forest management, tree/stand improvement, disease and insect control, and fire management.
5) Improvement of technology for procuring, processing, and manufacturing of forest products.
6) Global forest resources and forest product markets.
7) Advancements in science, technology, and other professional that could affect forestry profession."—Jianbang Gan, professor and coordinator, Forest Resources Program, and Walter A. Hill, dean, College of Agricultural, Environmental and Natural Sciences, Tuskegee University

"More emphasis . . . on the non-timber products . . . such as high quality water, sustained water supply, wildlife and recreation while at the same time seeking to preserve and enhance biodiversity . . . This will add to the challenge of finding ways to maximize production of wood and fiber."—David G. Topel, dean and director, College of Agriculture Experiment Station, Iowa State University

"The greatest need in forestry (and related resource) research and professional education is fundamental training of all specialists in systems management so they can both specialize and keep a perspective on how their specialty relates to all other specialties."—Chadwick D. Oliver, Forest Management and Engineering Division, College of Forest Resources, University of Washington

"The knowledge base . . . is somewhat lacking, especially with regard to non-market goods and services..." —Pete Morton, The Wilderness Society

"Expectations for rising population, rising standards of living, declining forest land base, and greater environmental awareness worldwide, call for even greater attention to providing new technology as a critical factor of production and a critical component of forest resource conservation." —Thomas E. Hamilton, director, USDA Forest Service, Forest Products Laboratory, and Ramsay Smith, Louisiana State University

"The ability to monitor as part of an adaptive management program should be explored." Virginia H. Dale, senior Scientist, Environmental Sciences Division, Oak Ridge National Laboratory

"Four research topics . . . essential to effective reserve design . . . include: 1) sometimes divergent implications of 'biodiversity' versus 'biotic integrity' in guiding forest policy, 2) assessing cumulative (and simultaneous) impacts of natural and anthropogenic impacts to forest function, 3) adding temporal scaling to spatial scaling as a factor in sustaining forest landscapes, and 4) the need to incorporate ecological measures of risk and uncertainty into forest planning." J. Christopher Haney, Ecology and Economics Research Department, The Wilderness Society

"Increasing the capacity of forestry research will require considerably more support for social science work and the study of the relationship between resource use and community capacity and well-being."—Jonathan Kusel, director, Forest Community Research

Input related to the essential knowledge base can be summarized as follows:

- *Universities.* Most respondents emphasized the need to teach students social sciences, wood-processing, research and technology transfer, "new forestry" and ecosystem approaches to management, and critical and multidisciplinary thinking and problem-solving. There was concern that the present knowledge base is not sufficient to address future forest research and management issues and that as faculty members retire, it is difficult to fill vacant positions in entomology, forest pathology, management, and forest products.
- *Government agencies.* Respondents recommended that forestry research and development focus on air and water quality, soil productivity, human use of resources, landscape fragmentation, population impacts, forest ecosystems, landscape ecology, use of latest technologies, and environmental effects of wood-fiber recycling and disposal of treated wood.
- *Trade organizations.* Three of the four respondents felt that research in social sciences (including rural sociology) is needed.
- *Nongovernmental organizations.* The knowledge base is lacking in nonmarket goods and services; many of the goods are produced by wildlands, so emphasis on wildlands economics should be increased.
- *Private industry and others.* Good knowledge of biology is needed, including forest ecology, insect and disease management and monitoring, tree breeding, and sustainable forestry.

During the National Research Council Workshop on Forestry Research Capacity, the question of whether the current knowledge base is adequate was addressed. The following seven major gaps in the knowledge base were identified in priority order:

1) Measurements, monitoring, and information systems.
2) Biologic knowledge base (biology and biochemistry, above- and in ground).
3) Management sciences (modeling, planning, forest-systems management, silviculture, agroforestry, restoration, and tree improvement).
4) Systems understanding (integration of biology, physical sciences, sociology, and risk management).
5) Forest health.
6) Human-natural resource interaction.
7) Wood science.

Additional gaps were noted but not ranked. These included, and reinforced, some previously noted gaps, such as

- Genetics (nine gaps proposed).
- Communication with the public and forest landowners (three gaps).
- Management options to favor biodiversity and productivity (or vice versa).
- More-efficient wood recycling and use.

- Models for long-term analysis of tradeoffs (biologic and economic), including spatial and climate-change considerations.
- Cumulative effect of forest management on the landscape scale.
- Better research planning among agencies.
- Monitoring.
- Fire suppression and management technology.
- Risk analysis.
- Spatial and temporal-scale implications.
- Economic models to address the international wood market.

Some new knowledge gaps identified during the workshop included (without ranking)

- Role of early succession species.
- Role of in-ground processes in sustaining ecosystems.
- Invasive species and ecosystem restoration.
- Management of carbon storage.
- "Backyard" silviculture.
- Chemical derivatives from wood.
- Mechanisms of ecophysiology—biologic changes underlying establishment and adaptations.
- Long-term effects of stream buffers in conifer forests on aquatic ecosystems.
- Composites—wood and other materials.
- Chemical treatment strategies for wood protection that are environmentally acceptable.
- Long-term impacts of chip-harvesting on forest ecosystems.
- Strategies for handling and dispersing forest recreation.

Addressing those gaps in the knowledge base in the short-term will require careful and thoughtful approaches to interdisciplinary research. In the longer term, it will require that our nation's education system attract and maintain high-quality students in programs designed to prepare competent research scientists to address the issues (see Chapter 4).

CONCLUSIONS AND RECOMMENDATIONS

Forestry education and research have long supported forest-management activities and principles that are fundamental and will always be considered essential; forest genetics is perhaps the best example. Forestry education and research have only recently begun to explore the interactions of social, economic, and environmental factors that are related to sustainable ecosystem management. The high-priority foundation science education and research fields related to forestry are biology, ecology, and silviculture; forest genetics; forest management, economics, and policy; and wood and materials science. The high-priority emerging education and research fields are human and natural resource interactions; ecosystem function, health, and management; forest systems on various scales of space and time; and forest monitoring, analysis, and adaptive management; and forest biotechnology.

Recommendation 2-1

To achieve an adequate knowledge base, forestry and natural-resource education and research programs in government and academia should dedicate resources to the foundation fields of forestry science while engaging in efforts to develop emerging education and research priority areas.

The high-priority fields can be addressed in an interdisciplinary manner with an appreciation of appropriate temporal and spatial scales. Addressing those fields will give policy-makers and managers the knowledge they need to implement a forest-management paradigm that will engender broad public and political support and meet society's physical and aesthetic needs.

The education and research priorities are aimed at the dual related goals of sustaining forests for a broad set of values, as recommended by the Committee of Scientists (1999), and providing the forest products required by a growing society. The present trend of a declining contribution of forest products from public forest lands and increasing recreation and nonmarket goods and services, and the trend of increasing intensification of forest-product output from private industrial forest lands, are likely to continue in meeting the diverse needs that society places on our forests. Foundation and emerging forestry education and research will be needed to provide professionals for our future to and support the policy and economic decisions that governments and the forest-product industry will need to make with respect to how and from where forest services and products are provided. It will be critical to retain the essential education and research on which the sustained delivery of forest products depends as resources continue to shift to sustainability related areas. That does not mean that all fields of forestry education and research need to be maintained at present levels. It does mean that current education and research need to be evaluated objectively and essential fields given high priority.

3

Current Forestry-Research Capacity in the United States

The nation's ability to provide adequate goods and services from forests, or even to maintain current area of forests, in the face of increasing population and consumption, is at risk. Improved protection of existing forests, afforestation of non-forested areas, reforestation after timber harvests, restoration of degraded forests, and increased productivity of new and existing forests—for commodities and noncommodity purposes—are required if demands for forests and for forest sustainability are to be balanced on the stand, landscape, or global scale. Research and monitoring underlie sustainable forest management and protection.

Scientific research is key to being able to identify how to improve forest conditions, allow compatible human uses, and sustain productivity for market and nonmarket goods and services. Research on forest products and use conducted by the USDA Forest Service, for example, has contributed to the development of knowledge and technology that have tripled the amount of fiber available for use from trees within the last 100 years (Lewis, 2000). Research on recycling of wood-based products has increased paper-recovery rates from 25 percent to 45 percent of fiber (Lewis, 2000). A specific example is the scientific advance in recycling of 33 billion stamps produced each year by the U.S. Postal Service as a result of research on pressure-sensitive adhesives, which had presented substantial problems in recycling (Lewis, 2000). Other research advances include the development of composite products and improvement in housing constructions.

Monitoring provides the means to measure whether forest conditions—from area extent to timber productivity to biodiversity to ecologic integrity—are being degraded, sustained, or enhanced. Monitoring provides the means for determining how the interaction of management interventions and natural climatic variations are affecting the forest resource, and suggests when new approaches are required. Such an integrated adaptive-management or systems approach to sustainable forest management will be necessary to meet future social needs and objectives.

Research and monitoring make it possible to determine how forests should be managed, including whether, how, and when intervention in natural conditions is needed. Research and monitoring are essential in the development of efficient approaches to developing intensive timber plantations, restoring degraded forests to better functioning ecologic systems, and providing the amenity and spiritual values that are sought by people.

ASSESSING FORESTRY-RESEARCH CAPACITY

Just as monitoring of forests is necessary to ensure future growth and sustainability, monitoring the status of forestry research is important to ensure future strength and capacity. The extent and condition of forests are uncertain; more importantly, the status of the nation's capacity to address these issues through forestry research is uncertain.

The capacity to achieve sustainability is highly variable and is positively correlated to the resources dedicated to forestry research (Szaro et al., 2000). It is possible to measure the input (human resources, financial resources, facilities, and equipment) into forestry research and its output (technology improvements, publications, economic development, and ecologic improvement), and a relatively thorough investigation of forestry research reveals greater capacity than perhaps widely recognized. However, how to focus and build that capacity are perhaps the most relevant questions for the next decade.

This chapter of the report summarizes available data on forestry-research capacity in terms of human resource, institutional, and financial inputs. We considered input and output to forestry research to describe the current status of the nation's forestry research environment, and to assess the adequacy of the nation's capacity to meet current and future needs. We also provide an overview describing evaluations of output (perceived return on investment). Where possible, we analyze the question of capacity in different disciplines; this was one of the specific concerns that prompted our study.

A PORTRAIT OF THE
FORESTRY-RESEARCH WORKFORCE

As described by Bengston (1998), the research capacity of a nation is determined in part by factors within the research system, such as the quantity and quality of resources available for research and characteristics of the institutional environment in which research is carried out. It is also influenced by national characteristics, including education systems, and public and private sector roles in research. To assess current U.S. forestry-research capacity, we review the primary forestry-research organizations here. To the extent possible, we describe the levels of manpower and research support they have provided currently and historically.

Research conducted by the U.S. Department of Agriculture (USDA) Forest Service is examined as a major contributor to the nation's forestry-research portfolio, as is research performed by forestry departments, schools, and colleges throughout the United States. Research related to forests in such departments and agencies as the U.S. Environmental Protection Agency (EPA), the National Aeronautics and Space Administration (NASA), the U.S. Department of Energy (DOE), the U.S. Department of Defense (DOD), the U.S. Department of Interior (DOI), and the National Science Foundation (NSF) is also germane. A direct research linkage to forests, at least where the links can be ascertained and quantified, is important in determining the status of forestry-research capacity. It would not include research in areas such as botany, rural sociology, or even sustainable agriculture, which although related, are more distant and more difficult to quantify.

USDA Forest Service

The USDA Forest Service Research and Development branch is the largest forestry-research organization in the world and is the largest contributor to the U.S. forestry-research workforce. It maintains 77 laboratories in 67 locations throughout the United States. They are organized within six regional research stations, a Forest Products Laboratory, and the International Institute of Tropical Forestry. Forest Service research is managed through regional research stations and each research station is made up of several Research Work Units (RWU's) located at Forestry Sciences Laboratories or on university campuses. RWU research is typically specialized in a particular subject area such as soil productivity, recreation, or forest insects. Each RWU typically conducts studies focused on its area of expertise or through interdisciplinary research projects that address complex problems of natural resource management and conservation. Interdisciplinary projects typically involve scientists from other work units, other parts of the Forest Service, other agencies, and universities. Forest Service trends in forestry research are by no means the only indicator of forestry-research capacity, but they provide accessible measures to obtain and track. Trends in Forest Service research funding, personnel, facilities, and Research Work Units (RWUs) are summarized in Tables 3-1 and 3-2.

Table 3-1 summarizes trends in the number of scientist years (SYs), RWUs, and research locations for Forest Service research. The agency had 964 SY equivalents in FY 1980 and pared that number to 633 by FY 1998. During the same period, the number of RWUs declined from 246 to 137— through both attrition of scientists and consolidation of RWUs to achieve greater administrative efficiency. The number of research locations dropped less precipitously, from 86 in FY 1980 to 67 in FY 1999. Although definitive data are lacking, it is commonly believed that Forest Service research infrastructure—the physical plant, equipment, and scientific technology—also declined in quality. Supportive of this belief is a report by an interagency working group on federal laboratory reform that released a report on improving federal laboratories in which the working group concludes:

"The (federal) laboratories' physical and human infrastructure is rich in capability but not fully matched to the challenges of the early twenty-first century." (National Science and Technology Council, 1999)

The working group report identifies the fact that each federal laboratory is important to its local and regional economy and employs people dedicated to national priorities. Examinations and review of infrastructure, capacity, and national needs have led to conclusions that there may be overcapacity in some parts of the federal system (National Science and Technology Council, 1999). Thus, when attempting to strengthen existing infrastructure, consideration must be given to weighing costs associated with maintaining facilities that may be obsolete and that may divert limited funds from more promising facilities.

Table 3-1. Forestry-Research Statistics for USDA Forest Service, FY 1980-2002.[a]

Fiscal Year	Appropriations, millions of $		Extramural Funding, millions of $			Scientist-Years[b] (FTE)	Research Locations	Research Work Units
	Actual	Constant 1980	Actual	Constant 1980	% Appropriations			
1980	111.5	111.5	10.6	10.6	9.5	964	86	248
1981	108.5	98.7	14.2	12.9	13.1	958	85	242
1982	112.1	95.3	10.8	9.1	9.5	908	83	235
1983	107.7	87.5	9.3	7.5	8.6	838	80	219
1984	109.4	85.6	7.7	6.0	7.0	813	77	207
1985	121.7	92.0	7.5	5.6	6.0	799	77	200
1986	120.1	88.4	10.4	7.6	8.6	734	78	199
1987	132.7	94.9	14.6	10.4	11.0	713	78	200
1988	135.5	93.6	18.3	12.6	13.5	724	76	190
1989	137.9	91.3	11.1	7.3	8.0	714	75	191
1990	144.7	92.0	13.2	8.4	9.1	716	75	190
1991	168.4	102.7	18.7	11.4	11.1	720	76	183
1992	181.3	107.4	29.6	17.5	16.3	714	78	183
1993	183.8	106.2	26.9	15.5	14.6	718	79	185
1994	193.1	108.9	21.5	12.1	11.1	720	78	185
1995	193.5	106.6	25.8	14.2	13.3	721	76	185
1996	178.0	96.1	14.7	7.9	8.2	692	69	185
1997	179.8	95.3	17.2	9.1	9.5	642	68	166
1998	187.8	98.4	17.6	9.2	9.3	633	67	137
1999	197.4	102.1	23.2	11.4	11.8	N/A	67	137
2000	217.7	104.3	21.6	10.3	9.9	841	N/A	137
2001	229.1	106.5	22	10.2	9.6	743	N/A	133
2002[c]	241.3	110.3	N/A	N/A	N/A	723	N/A	133

[a] Includes appropriated accounts only; excludes reimbursable accounts; [b] Scientist-year figures include term appointments of post-doctoral students. Actual numbers of permanent full-time researchers are lower by an estimated 25-50 FTEs for FY 1996-1999. For example, 606 permanent full-time researchers were employed in FY 1998 compared with 633 FTEs. 27 FTEs of effort were contributed by employees on term appointments in FY 1998; [c] Data for 2002 are not final.
Source: R. Guldin, USDA Forest Service, Washington, D.C., personal communication, July 1999. Drawn from Reports of the Forest Service, Fiscal Years 1980–1998; USDA Forest Service, Washington, DC, 2002 Budget Justification.

Table 3-2. USDA Forest Service Research Funding by Budget Line Item, FY 1980-2002 (thousands of $).[a]

Fiscal Year	Forest Protection	1980 $ Forest Protection	Resource Analysis	1980 $ Resource Analysis	Timber and Forest Mgmt	1980 $ Timber and Forest Mgmt	Forest Env. and Ecosystem	1980 $ Forest Env. and Ecosystem
1980	31,544	31,544	19,100	19,100	20,620	20,620	22,525	22,525
1981	29,883	27,089	18,347	16,631	20,705	18,769	32,133	29,128
1982	29,956	25,579	18,173	15,518	20,710	17,684	22,884	19,540
1983	30,061	24,870	17,316	14,326	20,585	17,030	21,813	18,046
1984	29,912	23,722	16,876	13,384	22,137	17,556	22,490	17,836
1985	29,110	22,292	21,646	16,577	22,161	16,971	22,421	17,170
1986	27,902	20,977	17,686	13,297	21,502	16,166	25,971	19,526
1987	31,224	22,648	22,218	16,116	23,891	17,329	30,580	22,181
1988	31,407	21,876	22,767	15,858	26,636	18,553	31,930	22,240
1989	32,944	21,892	22,636	15,042	27,383	18,197	33,912	22,535
1990	33,850	21,341	22,932	14,457	29,488	18,591	36,741	23,163
1991	38,168	23,091	25,807	15,613	36,550	22,112	43,373	26,240
1992	40,770	23,945	29,166	17,129	39,216	23,032	45,716	26,849
1993	40,833	23,285	30,720	17,518	39,594	22,578	46,033	26,250
1994	41,089	22,846	31,540	17,537	40,887	22,734	52,770	29,341
1995	36,998	20,004	32,361	17,497	52,924	28,615	43,083	23,294
1996	33,308	17,493	28,168	14,793	47,123	24,748	44,316	23,274
1997	33,559	17,229	26,341	13,523	50,284	25,816	45,369	23,292
1998	34,125	17,251	31,816	16,084	52,377	26,478	45,851	23,179
1999	34,307	16,968	39,021	19,300	50,664	25,058	48,924	24,198
2000	27,169	13,014	41,362	19,812	50,376	24,130	45,517	21,803
2001	29,934	13,919	37,530	17,451	53,536	25,824	50,406	23,439
2002[c]	30,363	13,876	38,044	17,386	55,631	25,423	51,453	23,514

Fiscal Year	Forest Products	1980 $ Forest Products	Subtotal	1980$ Subtotal	Other[b]	1980 $ Other[b]	Total	1980 $ Total
1980	17,742	17,742	111,531	111,531			111,531	111,531
1981	18,385	16,666	108,453	98,312			108,453	98,312
1982	20,422	17,438	112,145	95,759			112,145	95,759
1983	17,897	14,806	107,672	89,078			107,672	89,078
1984	17,988	14,266	109,403	86,764			109,403	86,764
1985	18,488	14,158	113,826	87,168	7,840	6,004	121,666	93,172
1986	17,560	13,202	110,621	83,167	6,506	4,891	117,127	88,058
1987	18,808	13,642	126,721	91,917	6,000	4,352	132,721	96,505
1988	19,770	13,770	132,510	92,297	3,000	2,090	135,510	94,387
1989	20,492	13,617	137,367	91,283	500	332	137,867	91,615
1990	21,142	13,329	144,153	90,881	500	315	144,653	91,196
1991	22,731	13,752	166,629	100,809	750	454	167,379	101,263
1992	25,640	15,059	180,508	106,014	750	440	181,258	106,455
1993	25,535	14,561	182,715	104,191	1,100	627	183,815	104,819
1994	25,697	14,288	191,983	106,744	1,100	612	193,083	107,356
1995	28,143	15,216	193,509	104,626			193,509	104,626
1996	25,085	13,174	178,000	93,482			178,000	93,482
1997	24,233	12,441	179,786	92,302			179,786	92,302
1998	23,775	12,019	187,944	95,009	(147)	74	187,797	94,935
1999	23,721	11,732	196,637	97,257	807	399	197,444	97,656
2000	22,310	10,690					186,734	89,449
2001	26,800	12,460					198,206	93,093
2002[c]	28,000	12,800					203,491	92,999

[a] Includes appropriated research only; excludes research construction and reimbursable accounts; [b] Includes funding for competitive forestry grants, challenge cost share, and congressional earmarks; [c] Data for 2002 are not final.
Source: Reports of the Forest Service, Fiscal Years 1980-1998; R. Guldin, USDA Forest Service, Washington, D.C., personal communication, October 1999; USDA Forest Service, Washington, DC, 2002 Budget Justification.

Research Scientists

Numbers of research scientists employed by the Forest Service are categorized by discipline for FY 1985-1999 in Table 3-3. As the table indicates, there has been a marked reduction in scientists in the agency from 985 in FY 1985 to 537 in FY 1999. In FY 1999, 136 (25 percent) of the research scientists were classified as foresters, 50 (9.3 percent) were classified as ecologists, 44 (8.2 percent) as wildlife biologists, and 31 (5.8 percent) as entomologists. The remaining 51 percent of the scientist work force was distributed among 31 employment classifications.

There has been a substantial shift in the classification of the Forest Service research scientists among disciplines. The greatest apparent reduction in expertise in the research branch is in the forester classification, from 350 in FY 1985 to 136 in FY 1999 (from 36 percent to 25 percent of the totals). Some of the reduction is not as much a proportional loss of expertise as an increase in specialization at the graduate level and an evolution of classification methods, but some silvicultural research positions and RWUs have been lost. The largest proportional loss of expertise has been in the forest products technologist classification, which dropped from 63 (6.4 percent of the total) in FY 1985 to 13 (2.4 percent) in FY 1999. Large personnel reductions also occurred in the job classifications for entomologists (70 to 31), plant pathologists (50 to 22), biologists (30 to 15), chemists (41 to 21), mathematic statisticians (30 to 12), soil scientists (27 to 15), range scientists (22 to 4), and mechanical engineers (14 to 3).

The largest increase in scientists was in the number of ecologists—from 9 in FY 1985 (0.9 percent of the total) to 50 (9.3 percent) in FY 1999. That probably reflects the increasing importance of ecology as a discipline over the last 15 years, the shift toward ecosystem management on federal lands, and the attractiveness of that research classification title to scientists. The only other groups that had more than a one-person increase were social scientists (9 to 14, offset by a 15 to 9 reduction in economists), and physical scientists (from 3 to 6).

In short, it is clear that Forest Service research capacity has decreased in terms of the number of scientists who are employed exclusively on a full-time permanent basis. The agency has hired many scientists on a temporary basis to work on major assessment projects, such as the President's plan and the Interior Columbia River Basin study. Those studies, however, tend to pull scientists away from basic research, and into applied, short-run data gathering, analysis, and synthesis projects. On balance, the substantial new assessment funds probably do little to build long-term research capacity.

The Forest Service also has hired an increasing number of persons with graduate degrees to work in the National Forest System and in state and private forestry. They might conduct modest studies and provide service to public land or private land managers, but they are not necessarily conducting long-term research relevant for the Forest Service. Again, there is probably not a net gain in applied research by employing persons with graduate degrees in other Forest Service branches, although the research knowledge obtained could be transferred more effectively by a larger complement of agency employees with graduate degrees.

Table 3-3. Number of Forest Service Research Scientists by Discipline, FY 1985–1988.

OPM Series	Title	1985	1988	1990	1995	1997[a]	1998[a]	1999[a]
101	Social scientist	9	7	8	17	12	13	14
110	Economist	15	11	11	11	9	6	9
150	Geographer	5	0	1	1	0	0	1
193	Archeologist	0	0	0	0	0	0	3
401	Biologist	30	16	13	14	13	14	15
403	Microbiologist	14	11	9	14	10	10	11
408	Ecologist	9	18	25	46	52	53	50
410	Civil engineer	6	3	1	0	0	0	0
414	Entomologist	70	62	55	38	35	30	31
430	Botanist	15	13	13	12	9	9	8
434	Plant pathologist	50	48	45	35	27	25	22
435	Plant physiologist	26	29	35	34	27	30	29
437	Horticultural	2	1	1	4	0	0	0
440	Geneticist	31	22	20	19	19	20	18
454	Range scientist	22	19	15	5	6	5	4
460	Forester	350	242	230	138	143	138	136
470	Soil scientist	27	27	28	19	17	16	15
482	Fishery biologist	8	8	11	14	11	14	14
486	Wildlife biologist	42	38	44	44	41	45	44
515	Ops. research analyst	7	1	2	0	0	0	0
801	General engineer	32	25	28	29	23	26	22
807	Landscape architect	1	0	0	0	0	0	0
808	Architect	1	0	0	0	0	0	0
810	Supvy res. civil engineer	0	0	0	0	0	0	1
819	Environmental engineer	0	0	0	0	0	0	1
830	Mechanical engineer	14	9	8	7	4	3	3
855	Electrical engineer	2	0	0	0	0	0	0
893	Chemical engineer	11	6	8	6	4	6	6
896	Industrial engineer	3	2	3	2	2	2	0
1301	Physical scientist	3	1	3	5	5	5	6
1310	Physicist	5	4	3	2	1	1	1
1315	Hydrologist	19	21	21	13	13	14	13
1320	Chemist	41	19	21	21	16	18	21
1340	Meteorologist	12	8	9	9	9	10	8
1350	Geologist	5	4	4	4	5	5	3
1380	Forest products technologist	63	43	31	25	21	18	13
1520	Mathematical	5	1	2	4	2	2	2
1529	Mathematical statistician	30	17	16	14	11	13	12
1530	Biological statistician	0	0	2	1	1	1	1
	Total	985	736	723	607	548	552	537

[a] Source: Nov. 22, 1996, Nov. 24, 1997; and Feb. 16, 1999; NFC Report, Count of Filled Positions Classified Under the RGEG.

Today, Forest Service scientists have a greater level of research support in terms of operating funds and support personnel than was the case two decades ago. Data in Table 3-1 show that the average budget in 1980 was about $116,000 per SY. By FY 2001, it had increased to about $308,000 per SY or $143,000 per SY in constant 1980 dollars. The average budget, therefore, has increased per SY, although the constant dollar total agency appropriations has declined to $106.5 million.

Research Productivity

Productivity or output measures have become increasingly important for government agencies in the last decade. Specifically, the Government Performance and Results Act (GPRA) of 1993 mandates that all federal agencies measure and report on the results of their activities annually. Agencies are required to develop a strategic plan that sets goals and objectives for a 5-year period and to produce an annual report of success in meeting them (Committee on Science, Engineering, and Public Policy [COSEPUP], 1999).

The GPRA process has prompted various efforts to define performance measures and collect information that can be used to track success. The National Academies have been examining means to implement the GPRA. A 1999 report (National Research Council, 1999, P. 9) suggested, as one of six major recommendations that:

> Federal agencies should use expert review to assess the quality of research they support, the relevance of that research to their mission, and the leadership of that research. Expert review must strive for having balance between having the most knowledgeable and the most independent individuals as members. Each agency should develop clear, explicit guidance with regard to structuring and employing expert review processes.

The Forest Service has collected data on research productivity for years before GPRA began and provided summaries on the productivity as measured by publications as part of this study on forestry-research capacity (Table 3-4). The data provided by the Forest Service summarize publications by aggregate budget line item in slightly different format from the budget data. The four broad categories of research were vegetation management and protection research (VMPR), wildlife, fish, watershed, and atmospheric sciences research (WFWAR), resource valuation and use research (RVUR), and inventory and monitoring research (IM). Table 3-4 shows the total reported publications summarized in the Forest Service research stations and RWU attainment reports, including internal publications by Forest Service scientists and external publications by cooperating scientists.

Scientists in the four broad categories of research had 1,886 publications in FY 1981, 2,299 in FY 1985, 3,021 in FY 1995, and 2,718 in FY 1998. Recall that the Forest Service (internal) scientist years for 1985, 1995, and 1998 were 985, 607, and 552 respectively. Thus the average number of publications was 3.06 per scientist in FY 1985, 5.0 in FY 1995, and 4.9 in FY 1998. Each of the four resource evaluation categories

Table 3-4. Number of Forest Service Publications by Discipline, FY 1981-1998.

Subject Area	RBAIS	1981	1982	1983	1984	1985	1986	1987	1988	1989	1990	1991	1992	1993	1994	1995	1996	1997	1998
Vegetation Management and Protection Research (VMPR)																			
Fundamental Plant Science	1.1	182	155	221	219	209	245	238	245	207	155	329	287	243	358	282	281	270	281
Silvicultural Applications	1.2	116	178	201	226	196	162	153	153	176	160	200	310	208	242	289	210	206	214
Quantitative Analysis	1.3	91	60	66	67	68	69	66	127	83	92	45	53	83	61	57	52	73	76
Forest and Rangeland Management	1.4	87	69	140	128	98	120	111	115	109	102	134	122	178	293	238	234	115	120
Forest Operations Engineering	1.5	39	38	50	66	84	71	70	57	40	46	50	73	58	71	58	49	59	61
Insects/Diseases/Exotic Weeds	1.7	406	447	440	431	489	428	411	339	328	383	337	403	427	480	383	364	279	290
Fire Science	1.9	86	78	105	65	102	88	86	113	56	84	100	100	75	114	101	99	112	116
Subtotal — VMPR	1.0	1,007	1,025	1,223	1,202	1,246	1,183	1,135	1,149	999	1,022	1,195	1,348	1,272	1,619	1,408	1,289	1,114	1,158
Wildlife, Fish, Watershed, and Atmospheric Sciences Research (WFWAR)																			
Terrestrial Wildlife Habitat	2.1	144	136	134	138	136	165	162	156	147	121	204	190	213	288	269	287	281	292
Aquatic Habitat	2.2	31	21	28	37	18	26	27	38	17	27	46	34	73	81	103	95	109	113
Watershed	2.3	149	141	183	119	189	177	173	215	219	292	141	219	181	226	301	282	253	263
Atmospheric Sciences	2.4	13	28	32	30	35	19	17	10	32	21	51	31	49	62	95	77	83	86
Subtotal — WFWAR	2.0	337	326	377	324	378	387	379	419	415	461	442	474	516	657	768	741	726	754
Resource Valuation and Use Research (RVUR)																			
Economics	3.1	94	122	128	142	182	205	196	131	190	159	142	215	168	200	175	187	113	117
Urban Forestry	3.2	33	23	41	25	36	45	42	31	17	58	46	2	49	60	40	51	37	38
Wilderness	3.3	7	6	9	6	7	7	6	4	5	8	9	23	8	11	9	15	16	17
Social/Cultural	3.4	64	54	78	53	62	59	56	40	49	74	77	211	68	97	78	135	144	150
Forest Product Utilization and Processing	3.5	212	170	221	210	192	197	188	102	144	142	157	169	238	244	285	258	240	249
Forest Product Safety/Human Health	3.6	44	63	66	67	65	84	80	126	71	72	81	101	70	80	64	59	108	112
Subtotal — RVUR	3.0	454	438	543	503	544	596	568	434	476	513	512	721	600	692	651	705	658	683
Inventory and Monitoring Research (I&M)																			
Forest Inventory & Analysis	4.1	88	92	99	119	110	143	138	203	109	120	107	123	105	122	102	166	78	81
Forest Health Monitoring	4.2	—	—	—	—	—	—	—	—	—	—	—	—	—	47	36	46	22	23
Monitoring Methods/Applications	4.3	—	—	—	—	—	—	—	—	—	—	—	—	23	—	—	—	18	19
Subtotal — I&M	4.0	88	92	99	119	110	143	138	203	109	120	107	123	128	169	138	212	118	123
General	0.0	—	28	17	31	21	21	20	22	79	49	148	7	20	71	56	58	—	—
GRAND TOTAL		1,886	1,909	2,259	2,179	2,299	2,330	2,240	2,227	2,078	2,165	2,404	2,673	2,536	3,208	3,021	3,005	2,616	2,718

Source: R. Guldin, USDA Forest Service, Washington, D.C., personal communication, July 1999.

increased their output of publications. WFWAR increased the most, from 337 in 1981 to 754 in 1999 (a 124% increase). IM publication numbers were fairly constant, VMPR increased about 15 percent, and RVUR increased rapidly and then declined to about a 50 percent increase over the base year, 1981.

Those trends appear to indicate that Forest Service researchers have become more productive in the measure that is most easily quantified. Some of that could be inherent productivity gains, some a response to fears that less productive RWUs and scientists will suffer reductions in force as budgets decline, and some gamesmanship in reporting to represent internal and external publications better. When productivity is evaluated in terms of the number of publications per year compared with the annual Forest Service research budget, it appears that productivity increased from approximately 25 publications per $1 million in 1985 to 28 publications per $1 million in 1998. Whether the Forest Service scientists and RWUs are actually more productive in their overall contributions to advancing the state of science or increasing knowledge remains moot.

Research Quality

Quality of research programs is more difficult to measure than financial resources and publications. With the pressure of increased productivity, Forest Service and other researchers are required to respond to the most quantifiable indicators of research success, which could potentially place too much emphasis on publications. That might harm research and shift efforts toward more applied or superficial topics and publication of "least publishable units" and away from challenging high-priority goals and seminal and integrative papers. The primary focus on applied or superficial topics also could adversely affect technology transfer efforts, in that they can receive less credit for research quality than other types of publications. The quality of research programs is hard to assess, as is their impact on forest management and protection. Such measures as success in receiving externally funded peer-reviewed grants or external peer reviews of science programs as suggested by the National Academies (1999), might be required to assess research program quality in the Forest Service and other forestry-research organizations.

Research Advisory Body

The Forest Research Advisory Council was authorized in 1995 and was reestablished by departmental regulation in 2002 as a requirement of the Agriculture and Food Act of 1981, Section 1441c to provide advice to the Secretary of Agriculture on accomplishing efficiently the purposes of the Act of October 10, 1962 (16 U.S.C. 582a et seq.), commonly known as the McIntire-Stennis Act. The Council provides advice related to the Forest Service research program and reports to the Secretary on regional and national planning and coordination of forestry research within the Federal and State agencies concerned with developing and utilizing the Nation's forest resources, forestry schools, and the forest industries. In addition, the Council provides advice to the Secretary on the apportionment of funds for the McIntire-Stennis Program. The Council consists of 20 members appointed by the Secretary. These members are drawn from federal, state, university, industry, and volunteer public organizations. Support to the Council is provided by the USDA Cooperative State Research, Education, and Extension Service and the Forest Service and it is served by 0.3 staff years.

The functions and responsibilities of the council include:

- Meeting at least once annually
- Reporting to the Secretary on regional and national planning and coordination of forestry research within the Federal and State agencies, forestry schools, and the forest industries
- Advising the Secretary on apportionment of funds
- Making special reports to the Secretary jointly through the Under Secretary for Research, Education, and Economics and the Under Secretary for Natural Resources and the Environment.

The Council has most of its membership coming from university and industry, and could be better balanced with perspectives needed to address the Council's charter. Needed perspectives other than those of the USDA Forest Service include a broader range of research partners and colleagues, stakeholders, users, and planners. The Council's work could be enhanced with input from more federal agencies outside of the USDA and the EPA, the only two federal agencies represented on the Council. Although the members of the Council work with others in the scientific community apart from the USDA and EPA, the council's work would benefit from broader perspectives offered by professionals in other government agencies, universities, and other research organizations.

The charter of the Council provides it with the authority to make recommendations on funding, planning and coordination of forestry research. The opportunity for greater involvement of all sectors concerned with forestry research exists. The Council's work could be more effective if it were better focused on the portions of its duties concerned with setting research priorities of McIntire-Stennis funding and monitoring accomplishments, and advising the Forest Service with research planning and priorities

Professional Forestry Schools and Colleges

A large amount of research is performed in schools and colleges. Faculties are drawn from an array of disciplines. They teach, perform research, and provide extension and professional services. Their total contribution to forestry research is substantial, probably equaling or exceeding that of the Forest Service. Some 48 universities have Society of American Foresters-accredited forestry curricula, and more than 60 universities or colleges have identifiable forestry and natural resources programs.

Faculty

Table 3-5 summarizes the trends in forestry faculty employment at 53 universities that have forestry programs and is derived from the USDA Handbook 305 (1994). As of the 1993–1994 academic year, there were 1,459 faculty listed in the handbook as being in the principal forestry, wildlife, fisheries, or natural resources departments. That constitutes a slight decline from the 1,503 listed for 1984–1985, but it is probably within the error of tabulation, given the expanding nature of forestry and natural resources departments. Many colleges and schools have added departments that contribute to forestry research and teaching capacity but are not included in the totals in Table 3-5. The South and the Lake States had slight declines in numbers of forestry faculty; the Midwest had a large decline. The Rocky Mountains and the West increased their numbers.

Data are not available on this, but most colleges and departments have split appointments between research and teaching and to a lesser extent, extension. If research accounted for about half the faculty full-time equivalents (FTEs), there might be about 700 faculty research FTEs. In the aggregate, the total faculty research FTEs in the United States are apt to be greater than the total Forest Service scientist FTEs. The teaching FTEs also contribute to research capacity, particularly in relation to their influences on graduate students. These interactions are discussed Chapter 4.

Forestry Extension

Forestry and natural-resources extension programs provide direct support for disseminating research findings to research users, such as nonindustrial private forest landowners, urban residents, production and environmental interest groups, natural-resource professionals, state and federal agencies, local governments, and policy-makers. Formal or informal extension efforts provide help to ensure that research results are used expeditiously.

The Renewable Resources Extension Act (RREA) provides federal funding for cooperative extension efforts at qualifying state universities and colleges. RREA has been authorized for budgets of up to $15 million per year, but appropriations have been much less. Funding started at $2 million in 1982, and was $3.2 million in FY 1999. State cooperative extension funding has also contributed to programs that have extension forestry specialists or regional or county agents. According to our calculations derived from the National Association of Professional Forestry Schools and Colleges (NAPFSC, 1999) report, forestry extension at member institutions accounts for about $20 million per year, including RREA funds. Thus, RREA funds are leveraged with state and county funding sources, at about a 9:1 ratio. However, state funds for extension appear to be declining due to budget cuts by 2001.

The United States has 9.9 million nonindustrial private forest landowners (Birch 1996), who own 49 percent of the nation's forest land and 58 percent of the nation's commercial timberland (Smith et al., 2001). Technology transfer is also needed for the even greater number of urban residents and for public land managers. The sum of $20 million per year indicates that technology transfer is much more modestly funded than research.

Table 3-5. Trends in Forestry Employment in Universities.

Region and Institution	Number of Faculty		
	1984–1985[a]	1986–1987[b]	1993–1994[c]
East			
University of Connecticut	8	5	6
University of Maine	27	29	24
University of Massachusetts	20	21	33
University of New Hampshire	15	15	16
Rutgers College	13	13	15
Cornell University	27	31	42
State University of New York-CESF	120	121	106
Pennsylvania State	45	43	38
University of Rhode Island	10	10	10
University of Vermont	25	25	29
Virginia Tech	53	56	58
West Virginia University	38	40	29
Subtotal, East	**401**	**409**	**406**
Lake States			
Michigan State University	25	27	21
University of Michigan	51	59	33
Michigan Technological University	15	13	21
University of Minnesota	43	44	47
University of Wisconsin	42	47	42
Subtotal, Lake States	**176**	**190**	**164**
Midwest			
University of Illinois	21	18	18
Southern Illinois University	13	12	11
Purdue University	32	31	24
Iowa State University	12	12	13
Kansas State University	4	4	4
University of Missouri	34	34	19
University of Nebraska	9	9	5
Ohio State University	19	19	16
Subtotal, Midwest	**144**	**139**	**110**
Rocky Mountains			
University of Arizona	37	40	37
Northern Arizona University	12	19	23
Colorado State University	20	16	21
University of Idaho	51	54	39
University of Montana	28	30	35
Utah State University	16	17	14
Subtotal, Rocky Mountains	**164**	**176**	**169**
West			
University of California-Berkeley	37	36	47
Humboldt State College	12	14	12
University of Alaska-Fairbanks	7	6	7

Table 3-5. Trends in Forestry Employment in Universities. (continued)

Region and Institution	Number of Faculty		
	1984–1985[a]	1986–1987[b]	1993–1994[c]
University of Nevada	8	6	8
Oregon State University	81	72	75
Washington State University	26	28	39
University of Washington	58	55	54
Subtotal, West	**229**	**217**	**242**
South			
Auburn University	36	38	47
Alabama A&M University	2	2	6
University of Arkansas-Monticello	15	17	20
University of Florida	31	25	26
University of Georgia	36	38	34
University of Kentucky	12	14	15
Louisiana State University	28	28	19
Louisiana Tech University	10	8	9
Mississippi State University	27	26	30
North Carolina State University	81	81	73
Oklahoma State University	14	12	14
Clemson University	35	35	22
University of Tennessee	24	23	16
Texas A & M University	17	18	17
Stephen F. Austin College	21	22	20
Subtotal, South	**389**	**387**	**368**
Grand Total	**1503**	**1518**	**1459**
(Percentage of 1984–1985)	(100%)	(101%)	(97%)

[a] Source: *USDA Cooperative States Research Service Agricultural Handbook No. 305*: 1984–1985 Directory of Professional Workers in State Agricultural Experiment Stations and Other Cooperating State Institutions. January 1985.

[b] Source: *USDA Cooperative States Research Service Agricultural Handbook No. 305*: 1986–1987 Directory of Professional Workers in State Agricultural Experiment Stations and Other Cooperating State Institutions. January 1987.

[c] Source: *USDA Cooperative States Research Service Agricultural Handbook No. 305*: 1993–1994 Directory of Professional Workers in State Agricultural Experiment Stations and Other Cooperating State Institutions. January 1994.

A serious disconnect between forestry research and its application on the ground limits the application of existing and new knowledge (Callaham, 1989; NAPFSC, 1999). Universities, governments, and private companies share the responsibility for training technical professionals to function at the interface between science and management. In addition to training people to operate at the interface, forest managers need to be lifelong learners. Researchers need to listen and respond to forest managers' needs and to articulate the practical significance and benefits of their research. Extension personnel must play a critical role in transferring knowledge gained through research to applications in forest management.

Private Industry

It is estimated that one to several hundred scientific research personnel are employed in the forest industry. Industry research can be categorized into five areas: forest health, water quality, fish and wildlife, ecosystem management, and timber productivity. A large portion of the environmental research (research in categories other than timber productivity) focuses on environmental protection in the context of timber management, rather than on basic studies of flora and fauna. Most of the results of the environmental research are available to the public and the general scientific community, whereas most of the results of the timber-productivity research are not.

Total Forestry Research Workforce by Sector, Function, and Sustainable Forest Management Criteria

A recent survey conducted by the USDA Forest Service (2002) provides the most comprehensive and up-to-date estimate of the total forestry research workforce as of 2001. This survey was conducted in part, in response to early requests for input and data for this NRC study and in conjunction with Forest Service efforts to measure and monitor the U.S. participation in meeting the Montreal Process Criteria and Indicators, which are discussed in Chapter 2. The study surveyed all U.S. professional forestry schools and colleges, the Forest Service, and the U.S. forest products industry to determine personnel efforts in research, education, and extension, broken down into the seven criteria for sustainable forest management. This study provides a thorough summary of current comparative efforts for forestry research and development, except for the newer federal agencies, state organizations, and nongovernment organizations that perform forestry research. Table 3-6 summarizes the results.

The total effort reported by the Forest Service (2002) survey includes 1,346 full-time equivalents (FTEs) for all three sectors in research; 598 FTEs in teaching; and 243 FTEs in extension. In total, 2186 FTEs are devoted to research (62%), teaching (27%), or extension (11%) activities by the identified forestry organizations. Universities have the largest number of FTEs devoted to all three functions, with 1361 persons (62%), the Forest Service has the second largest workforce with 701 scientists (32%), and the forest industry has 124 forest scientist FTEs (6%). The Forest Service has the largest number of research scientist FTEs with 658 (49%), the universities follow with 575 (43%), and the industry has 112 (8%). Academia has 98.7% of the teaching FTEs (596), followed distantly by the Forest Service and industry with only one each. The university sector also has 78% of the extension FTEs (190), followed by the Forest Service with 42 (17%), and the private forest industry with 11 (5%).

These data provide a basis for comparison of earlier estimates. The previous estimate for total Forest Service personnel for 1999 was 537 research scientists. The Forest Service (2002) total of 658 FTEs shows an increase in the number of scientists over 1999, but this number probably includes post-doctoral positions and other scientists that may not have been included in the earlier data set. However, there has been an increase in Forest Service research funding and capacity in the last few years. On the other hand, the new Forest Service (2002) survey of university faculty in forestry departments found there were only 1361 FTEs, compared to 1459 by the USDA Handbook 305 (1994). The USDA Handbook includes forestry, fisheries, and natural resource departments. That broader definition probably accounts for the larger number of FTEs, and suggests that there may still be a fairly stable or even increasing capacity in all of the natural resource faculty at universities. There were no previous estimates of the forestry extension workforce. The 243 FTEs in the United States represents a substantial workforce, with at least some representation in all sectors.

The reported forest industry research workforce estimate was smaller than previously estimated, at 124 persons rather than several hundred. It should be noted that dozens to perhaps more than 100 scientists are involved in forestry research through state organizations, federal environmental agencies, and domestic and international environmental nongovernment organizations.

The data on effort by SFM Criteria demonstrate that for all research, education, and extension FTEs, Criterion 1 (biological diversity) and 2 (productive capacity) each included about 21% of the total scientists' effort, at about 450 FTEs. Criterion 6 (socio-economics) included 393 FTEs (18%). Ecosystem health and soil and water each had about 300 FTEs (14% each). The institutional framework and carbon cycles criteria had the smallest reported effort, with 166 FTEs (8%) and 122 FTEs (6%), respectively.

For research FTEs alone, the percentage effort among SFM Criteria was fairly similar. The research FTE effort by universities was the greatest in biological diversity criterion (155 FTEs), followed by socio-economics (131 FTEs), productive capacity (85 FTEs), and institutional framework (72 FTEs). The Forest Service research efforts were dominated by the productive capacity (158 FTEs), ecosystem health (156 FTEs), and biological diversity (112 FTEs) criteria. Forest industry research efforts were dominated with activity in productive capacity (67 FTEs) and soil and water (20 FTEs).

These findings indicate that universities appear to have the most diverse research portfolio and are relatively strong in social science efforts. The Forest Service focuses on core biological criteria, and appears to have a slight plurality of its research FTEs focused on productive capacity. The combined biological diversity and ecosystem health criteria workforce is effectively the largest area, and is focused on broader issues. The forest industry focuses on productivity questions, with soil and water quality also being important.

Table 3-6. Full Time Equivalents of U.S. Forestry Scientists by Sector, Function, and SFM Criterion, 2001.

	Sustainable Forest Management Criterion								
	1: Biological Diversity	2: Productive Capacity	3: Ecosystem Health	4: Soil and Water	5: Carbon Cycles	6: Socio-economics	7: Institutional Framework	Total	% By Function
Academic Institutions									
Teaching	155	85	50	77	28	131	72	596	44
Research	136	96	53	84	47	114	45	575	42
Extension	27	40	25	26	3	48	22	190	14
Subtotal	318	221	128	186	77	293	138	1361	100
% By Criterion	23	16	9	14	6	22	10	100	
USDA Forest Service									
Teaching	0	0	0	0	0	0	0	1	0
Research	112	158	156	86	41	80	25	658	94
Extension	9	3	10	6	1	10	3	42	6
Subtotal	122	161	166	92	43	90	27	701	100
% By Criterion	17	23	24	13	6	13	4	100	
Forest Industry									
Teaching	0	0	0	0	0	1	0	1	1
Research	10	67	5	20	3	9	0	112	91
Extension	1	8	1	2	0	0	0	11	9
Subtotal	10	75	5	22	3	10	0	124	100
% By Criterion	8	60	4	17	2	8	0	100	
All Sectors									
Teaching	155	85	50	77	28	132	72	598	27
Research	258	321	214	189	91	203	70	1346	62
Extension	37	51	35	34	4	59	24	243	11
Total All Functions	450	457	299	300	122	393	166	2186	100
% by Criterion	21	21	14	14	6	18	8	100	
Total Research Only	258	321	214	189	91	203	70	1346	
% by Criterion	19	24	16	14	7	15	5	100	

Source: USDA Forest Service 2002
Notes: Data may not add exactly due to rounding

INVESTMENT IN FORESTRY RESEARCH

Focusing on the financial resources devoted to forestry research overlooks other important factors in measuring research capacity, such as human resources devoted to research. Although research funding is often used as a proxy for these inputs and for research activity in general, it is important to note that research funds purchase the services of scientists and research personnel and the equipment they use.

Forest Service Research Support

Support for Forest Service research encompasses several components, including direct research appropriations, construction appropriations, and reimbursable expenses. Trends for appropriated Forest Service research dollars are summarized in Table 3-1. Research received $111.5 million in FY 1980 and $229.1 million in FY 2001. If the cost of inflation is accounted for (converting the funding to constant 1980 dollars), Forest Service research funding dropped from $111.5 million in FY 1980 to $106.5 million in FY 2001. The total appropriated budget for Forest Service research in FY 2001 was also lower (in constant 1980 dollars) than the FY 1994 budget, which had $200 million in appropriated funds and $19.6 million in reimbursable accounts, for a total of almost $220 million or $122.3 million in constant 1980 dollars. In addition, the agency periodically has received congressional earmarks for specific projects, locations, or buildings, particularly in the late 1980s, which reduce the discretionary budget even more than is apparent by looking at total funding.

Forest Service research funds were appropriated by broad disciplinary budget line items (BLIs) until FY 1994 and have since been consolidated into one appropriation. Table 3-2 summarizes these appropriations by BLI from, with estimated breakdowns being made for the last five fiscal years. FY 1994 was the last year that Forest Service research received direct funding for major BLIs, rather than as one consolidated research budget. At that time, research programs were divided among forest protection (about $41 million), resource analysis ($32 million), forest management ($41 million), forest environment and ecosystem ($53 million), and forest products and harvesting ($26 million).

The trends in Forest Service research vary among disciplines. Forest-protection research funding decreased from $32 million in FY 1980 to $14 million in constant 1980 dollars or $30 million in 2001 dollars. Forest products research had an increase of about $9 million in funding between 1980 and 2001, which is actually a decrease of $5.2 million if inflation is considered. Resource analysis experienced an increase, with FY 2000 in constant 1980 dollars slightly higher than the FY 1980 level, a $712,000 increase. Timber and forest management increased the most, by about $5 million in constant 1980 dollars, and the forest environment and ecosystem BLI increased by about $1 million in constant 1980 dollars.

The appropriations statistics in Tables 3-1 and 3-2 indicate that in constant dollars, changes in appropriations appear to have been favorable to forest-industry related research (i.e., timber and forest management research). Research areas that focus on the environment (i.e., forest environment and ecosystem research) did not fare as well. In general, while appropriations fell by 8.4 percent in constant dollars in the two decades following 1980, they went up by 27.0 percent in timber and forest management research. They fell by 43.8 percent in forest protection (forest health) research and went up by only 12.3 percent in forest environment and ecosystem research despite the increased public attention to issues in both of these areas. The increase in forest timber and forest management research appropriations may surprise some, since the forest industry generally perceives that the Forest Service does substantially less timber productivity research.

The 1990 National Research Council report on forestry research called for a reorientation of research (and forest management) to address environmental concerns. If the recommendations of the 1990 report had been implemented, environmental and ecosystem research appropriations presumably would have experienced a greater increase, but in fact these areas remained relatively flat in real terms since the early 1990s, while timber and forest management research went up by more than 50 percent in constant dollars. However, strict interpretation of these data must be qualified to recognize that forest-industry research priorities have broadened in the past decades to include environmental forestry and performance, water issues, and sustainable productivity (AFPA, 1996). Furthermore, the Forest Service shifted to a broader set of research projects in the forest management category than was performed in the old timber management category. The appropriations categories are broad and may conceal actual research priorities, but, if assessed literally, the numbers appear to reflect Forest Service and Congressional priorities.

The Forest Service provides extramural research contracts, grants, and cooperative agreements to universities, nonprofit organizations, and some other private organizations (Table 3-1). Extramural funding has varied considerably since FY 1980, when the Forest Service provided $10.6 million to cooperating organizations. Extramural research funding peaked in FY 1992, when the Forest Service provided $17.5 million in constant 1980 dollars ($29.6 million in 1992 dollars); in FY 1996, it decreased to 7.9 million in constant 1980 dollars ($14.7 million in 1996 dollars), and in FY 2001 it increased to $10.2 million in constant 1980 dollars ($22 million in 2001).

In FY 1981, the Forest Service spent $14.2 million of $108.5 million on extramural contracts, grants, and cooperative agreements. That was 13.1 percent of its total research budget. In FY 1992, the agency spent approximately 16.3 percent on extramural research, and this declined to 9.4 percent in FY 1998. While data are lacking, it appears that almost all of these agreements are negotiated among individual Forest Service scientists and university professors, lacking any broad competition or peer review. The agency does not sponsor any national competitive grant research.

Forest Service extramural funding went mostly to land grant universities in FY 1997 — $15,360,000 of the $19.9 million total. However, in FY 1998, non-land-grant universities received the largest amount of extramural research funding — $7,654,000 compared with $7,595,000 for land-grant institutions. The 1890 historically black colleges and universities received about a half million dollars in extramural funding in both fiscal years. Nonprofit organizations received about $1.6 million; state and local governments, $250,000-825,000; and foreign for-profit and nonprofit organizations and private individuals, about $85,000 (USDA Forest Service, 1999).

Other Federal Forestry-Research Funding

The largest other direct federal funding for forestry research is provided under the authorization of the McIntire-Stennis Act of 1962. The act provides federal financial support to colleges and universities for forestry research and graduate education. Whereas Forest Service research budgets are authorized and appropriated in Congress in conjunction with those of the Department of the Interior and related agencies, McIntire-Stennis funds are appropriated in the congressional agriculture committees, similar to the Hatch Act funds for the state land grant agricultural experiment stations.

Table 3-7 summarizes federal appropriations for McIntire-Stennis funds from 1980 to 1999. They are adjusted to constant 1980 dollars for comparison of purchasing power. The McIntire-Stennis expenditures increased gradually from $9.7 million in FY 1980 to $11.9 million in FY 1987, jumped to $16.8 million in 1988, and increased to $21.9 million by FY 2000. The funds in actual dollars have increased only slightly in constant 1980 dollars.

Three major factors have been used to determine the proportion of McIntire-Stennis funds that states receive (National Research Council, 1990, P. 18):

- Proportion of acreage in commercial forest land (40%)
- Volume of roundwood produced (40%)
- Amount of nonfederal money spent on forestry research (20%)

Some flexibility is built into the formula; for example, the weight of the factors is not mandated by law, but is set by the Secretary of Agriculture.

Table 3-8 summarizes the McIntire-Stennis appropriations for FY 2000. The funds are distributed among the public forestry schools and colleges throughout the nation, and they provide crucial support for many forestry programs. Fifteen of the 50 states each received more than $500,000 in 2000, and each state received more than $50,000. States with more than one forestry school or college split the funding among relevant institutions. Once allocated to states, funds may be administered or allocated entirely within the relevant forestry school, college, division, or department, or they may be distributed competitively among the faculty at a university (Box 3-1). Proposals have been made to extend McIntire-Stennis funding to other institutions, such as 1890 historically black colleges and universities and 1994 Native American natural-resources schools. Such extensions would require additional funding, and additional faculty and infrastructure at recipient institutions, to enhance national research capacity.

The McIntire-Stennis funds provide a foundation for maintaining forestry research at qualifying institutions. This relatively reliable source of funds has undoubtedly allowed the expansion of university forestry-research capacity throughout the country. It also has ensured that all qualifying schools receive some funds each year and that funds are well distributed geographically, institutionally, and programmatically. Even small forestry programs are able to perform some applied research with McIntire-Stennis funds, as well as to support graduate education. Funding under the act can be compared with the competitive grant approach, where the large research institutions tend to receive by far the most funds and have the most success, and small schools are rarely able to compete well. The formula funds provide a base level of support for forestry research and graduate education at qualifying public forestry schools, colleges, and departments throughout the nation. They are particularly important for providing reasonable stability, especially for long-term research, which is important to adequately address foundation and emerging issues (University of Idaho, 1983). McIntire-Stennis funds may also be spent on forestry research in other departments or colleges, depending on agreements in individual states.

Another strength attributed to the availability of research funding through formula allocation like McIntire-Stennis is that it reduces the proportion of a researcher's time spent applying for competitive grants. Because grant application processes are time consuming and can have low rates of success, formula funding allows more time for researchers to devote to performing the funded research (Huffman and Evenson, 1993).

In recent years, formula funds, as McIntire-Stennis and Hatch Act appropriations are referred to, have been considered less desirable than competitive grants. Concurrently the forestry-related formula allocations have received relatively small increases, which has resulted in a reduced the share of federal research funds appropriated through USDA compared with other federal agencies, such as NASA, EPA, and NSF. The small increases in formula funding might be attributed to decline in political influence of rural agricultural and forestry interests, but a portion could be due to the nature of the administration of formula funds and their perception. One of the perceived weaknesses of formula funds is that research conducted with formula funds is not automatically subject to peer review.

Box 3-1
Hatch and McIntire-Stennis Proposals at the
College of Agriculture and Life Sciences, University of Wisconsin

Hatch and McIntire-Stennis (M-S) funding is open to faculty members in CALS, SOHE, and AHABS. Faculty from other colleges and universities may be collaborators on a project. Investigators may submit proposals for an individual-investigator grant, or a multiple-investigator interdisciplinary grant. The Hatch and M-S competition supports a wide range of research. While graduate training is central to use of formula funds, and encouraged as a typical request, some exceptions may be possible. Each proposal is judged on appropriateness of proposed research for formula funding, quality of the science, and likelihood of successful achievement of those goals. Interdisciplinary proposals with multiple investigators are considered in the open competition with the following considerations:

- High quality of research work proposed
- Special emphasis on problem solving for Wisconsin
- Realistic budgets and work specification
- Evidence that the interdisciplinary team has worked together on the proposal
- Plans to link the research to extension or outreach activities
- Demonstration of productivity from past and present formula funding for all collaborators

There are valid criticisms of the way peer review tends to operate nevertheless, many see peer review as the key to quality control in scientific endeavor (Chubin and Hackett, 1990; National Research Council 2000). Formula funds also have lacked effective means of accountability in terms of how they have been used by state institutions or whether they have been devoted to research issues that justify federal support (Alston and Pardey, 1996). Formula funds for forestry research are also provided under the Hatch Act, as part of the general federal support for the state land grant agricultural experiment stations. A modest amount of the Hatch funds is allocated to forestry research. The 1999 NAPFSC report estimated the total at about $2 million.

McIntire-Stennis and Hatch Act program reviews may be requested by cooperating institutions but are not required on a regular schedule. Perhaps a return to more regular external scientific reviews, such as suggested by the National Research Council (1999), would be desirable. Providing more research oversight and evaluation might be a way to ensure the quality of research conducted with these funds (University of California, 2001; Box 3-2). Greater competition for formula funds within schools also might broaden the base of scientists who perform research with formula funds and enhance quality. More external peer reviews and more funding competition will improve consistency and quality among formula funded research. Those steps also might foster better communication about and support for the programs.

Leveraging Research Support

The National Association of Professional Forestry Schools and Colleges (NAPFSC, 1999) provides a detailed history and status report of the McIntire-Stennis program and its accomplishments. Many of the successes of the program and its effectiveness in leveraging state and private-sector funds for forestry research are noted in the report. For FY 1997, total forestry research, extension, and education funding at public colleges and universities was about $204 million (NAPFSC, 1999, P. 7):

> McIntire-Stennis funding represents 10 percent of the $204 million used by American forestry schools for research, education, and extension. Other federal funds are about 24 percent, States provide 44 percent, industry contributes just over 7 percent, and other non-federal sources (e.g., foundations) add about 14 percent. In 1997, the total federal funding for forestry research, education, and extension at forestry schools was $70 million—about 34 percent of the total. Federal dollars are awarded through formula, competition, and cooperative agreements to achieve goals of national importance. Each McIntire-Stennis dollar leverages approximately nine dollars of other federal, state, and private sources.

The southern NAPFSC group collects data that allow examination of McIntire-Stennis funding compared with other sources. In FY 1998, the 14 reporting southern forestry land grant schools, colleges, and departments had total research budgets of $36 million.

Table 3-7. McIntire-Stennis Funding in Actual and Constant Dollars, FY 1980-2000.

Fiscal Year	Appropriations, millions of $	
	Actual	Constant 1980
1980	9.7	9.7
1981	10.4	9.5
1982	11.3	9.6
1983	11.8	9.6
1984	12.2	9.5
1985	12.4	9.4
1986	11.9	8.8
1987	11.9	8.5
1988	16.8	11.6
1989	17.1	11.3
1990	16.6	10.6
1991	17.1	10.4
1992	17.7	10.5
1993	17.7	10.2
1994	19.8	11.2
1995	19.8	10.9
1996	19.4	10.5
1997	19.4	10.3
1998	20.5	10.7
1999	21.9	11.3
2000	21.9	11.1

Sources: 1) For FY 1980-1997, summary data of McIntire-Stennis Program expenditures as reported on form AD419 to USDA Current Research Information System by recipient institutions.
2) For FY 1998 and 1999, as reported on USDA/CSREES form OD-1088-D, March 5, 1998, and February 9, 1999, respectively.
3) For FY 2000, Agricultural Appropriations Act: FY 2000.

Table 3-8. Distribution of McIntire-Stennis Funds to Eligible State Institutions or Institutional Units.

Location	Institution	Amount, $
Alabama, Auburn	Auburn University	716,214
Alaska, Fairbanks	University of Alaska	446,158
Arizona, Flagstaff	Northern Arizona University	144,704
Arizona, Tucson	University of Arizona	147,849
Arkansas, Fayetteville	Agricultural Experiment Station, University of Arkansas	605,514
California, San Luis Obispo	California Polytechnic State University	97,538
California, Arcata	California State University, Humboldt	93,154
California, Berkeley	University of California	455,176
Colorado, Fort Collins	Colorado State University	337,309
Connecticut, New Haven	Connecticut Agricultural Experiment Station	171,346
Connecticut, Storrs	Storrs Agricultural Experiment Station, University of Connecticut	57,115
Delaware, Newark	University of Delaware, Agricultural Experiment Station	65,185
Florida, Gainesville	Agricultural Experiment Station, University of Florida	553,581
Georgia, Athens	School of Forest Resources, University of Georgia	731,899
Guam, Agana	University of Guam	34,311
Hawaii, Honolulu	University of Hawaii	167,268
Idaho, Moscow	University of Idaho	469,001
Illinois, Carbondale	Southern Illinois University	156,014
Illinois, Urbana	University of Illinois	156,014
Indiana, Lafayette	Purdue University	385,716
Iowa, Ames	Agriculture and Home Economics Station, Iowa State University	269,279
Kansas, Manhattan	Kansas State University	133,216
Kentucky, Lexington	Agricultural Experiment Station, University of Kentucky	418,948
Louisiana, Baton Rouge	Louisiana State University, School of Forestry	439,801
Louisiana, Ruston	School of Forestry, Louisiana Tech University	190,994
Maine, Orono	University of Maine	568,614
Maryland, College Park	University of Maryland	242,065
Massachusetts, Amherst	University of Massachusetts	300,406
Michigan, East Lansing	Michigan State University	207,679
Michigan, Houghton	Michigan Technological University	207,679
Michigan, Ann Arbor	University of Michigan	207,679
Minnesota, St. Paul	University of Minnesota	511,998
Mississippi	Mississippi State University	677,464
Missouri, Columbia	School of Forestry, University of Missouri	459,756
Montana, Missoula	University of Montana, Forestry and Conservation Experimental Station	439,444
Nebraska, Lincoln	University of Nebraska	171,870
Nevada, Reno	University of Nevada, Mac C. Fleishmann College of Agriculture	116,010
New Hampshire, Durham	University of New Hampshire	282,884
New Jersey, New Brunswick	Rutgers State University, Agricultural Experiment Station	201,247
New Mexico, Las Cruces	New Mexico State University	255,490

Table 3-8. Distribution of McIntire-Stennis Funds to Eligible State Institutions or Institutional Units, FY 2000. (continued)

Location	Institution	Amount, $
New York, Ithaca	New York State College of Agriculture and Life Sciences, Cornell University	168,754
New York, Syracuse	State University of New York, College of Environmental Sciences	497,895
North Carolina, Raleigh	North Carolina State University	696,688
North Dakota, Fargo	North Dakota State University of Agriculture and Applied Sciences	106,003
Ohio, Wooster	Ohio Agricultural Research and Development Center	391,734
Oklahoma, Stillwater	Oklahoma State University	364,522
Oregon, Corvallis	Oregon State University	745,495
Pennsylvania, University Park	Agricultural Experiment Station, Pennsylvania State University	514,999
Puerto Rico, Rio Pedras	Agricultural Experiment Station, University of Puerto Rico	92,397
Rhode Island, Kingston	University of Rhode Island	80,165
South Carolina, Clemson	College of Forestry and Recreation Resources, Clemson University	541,402
South Dakota, Brookings	South Dakota State University	139,677
Tennessee, Knoxville	University of Tennessee	500,665
Texas, Nacogdoches	Stephen F. Austin State University	283,019
Texas, College Station	Texas Agricultural Experiment Station, Texas A&M University	299,659
Utah, Logan	Utah State Agricultural Experiment Station	160,428
Vermont, Burlington	University of Vermont	338,143
Virgin Islands, St Thomas	The College of the Virgin Islands	51,579
Virginia, Blacksburg	Virginia Polytechnic Institute and State University	582,220
Washington, Seattle	University of Washington	323,226
Washington, Pullman	Washington State University	323,227
West Virginia, Morgantown	West Virginia State University	405,340
Wisconsin, Madison	Agricultural Experiment Station, University of Wisconsin	486,977
Wyoming, Laramie	University of Wyoming	214,854
Total Payments to States		20,688,273
Federal Administration[a]		636,853
Small Business Set-Aside[a]		514,832
Biotechology Risk Assessment[a]		8,828
GRAND TOTAL		21,848,786

[a] Based on 1999 dollars converted into 2000 dollars.
Source: Allen Moore, USDA/CSREES Current Research Information System, Washington, D.C., personal communication, April 2002.

McIntire-Stennis regional shares totaled $5.2 million (14.4 percent). In comparison, the reporting schools had state appropriated research budgets of $13.3 million (37.0 percent); other federal, state, or private research grants of $14.0 million (38.9 percent); and other sources of income (timber sales, private industry contributions) of $3.6 million (10.2 percent). Thus, in the South, McIntire-Stennis funds, which help to provide base-level programmatic support, were leveraged at a 7:1 ratio from other research sources.

Box 3-2
Reviews Improve Quality of Forestry Research at the University of California Berkeley

The McIntire-Stennis Cooperative Forestry Program provides federal funds for forestry research at various universities throughout the country. In California, the University of California, Humboldt State University, and Cal Poly San Luis Obispo receive funding from the McIntire-Stennis Act, according to a formula set by the director of the California Department of Forestry and Fire Protection. Based on the Administrative Manual for the McIntire-Stennis Cooperative Forestry Program:

The scope of forestry research which may be conducted under the McIntire-Stennis (M-S) Act includes investigations relating to:
- reforestation and management of land for the production of timber and other related products of the forest
- management of forest and related watershed lands to improve conditions of water flow and to protect resources against floods and erosion
- management of forest and related rangeland for production of forage for domestic livestock and game and improvement of food and habitat for wildlife
- management of forest lands for outdoor recreation
- protection of forest and resources against fire, insects, diseases, or other destructive agents;
- utilization of wood and other forest products
- development of sound policies for the management of forest lands and the harvesting and marketing of forest products
- such other studies as may be necessary to obtain the fullest and most effective use of forest resources (Source: 16 U.S.C. 582a-6; USDA Forest Service, 1993)

Under the current College of Natural Resources (CNR) administrative structure, the overall research program of every faculty member is reviewed regularly for relevance to CNR and the systemwide Division of Agriculture and Natural Resources (DANR) missions, excellence of science, and quality of future research plans. This review is conducted by the CNR faculty Research Committee (RESCOM). All funds designated for support of faculty research, including McIntire-Stennis funds, are then allocated by the Dean to faculty members' projects according to the rating received from the RESCOM. This new review program has improved the allocation of these research funds by putting them on a merit basis within CNR. Following federal guidelines for all Federal Formula Funds, McIntire-Stennis funds are allocated only to those faculty members who have active McIntire-Stennis projects.
Source: University of California, Berkeley, http://www.cnr.berkeley.edu/forestry/macsten.html

University Research Support

Public and private funding for all university research has increased dramatically over the last 4 decades, from just under $3 billion in 1959 to an estimated $25 billion or more in 2000 (Committee for Economic Development, 1996). The preceding discussion of McIntire-Stennis funding described the broad distribution of funding sources for the $204 million for public forestry schools and colleges in FY 1997 for research, education, and extension (NAPFSC, 1999). State sources made up 44 percent of the total, other federal sources 18 percent, McIntire-Stennis 10 percent, and other non-federal sources 10 percent. Competitive grants and cooperative agreements and industry programs each constituted 7 percent ($14 million) of the $204 million total. The balance was comprised of self-generated income (4 percent), grants (3 percent), Renewable Resource Extension Act funds (2 percent), and Hatch Act funds (1 percent).

State funds support research, education, and extension. National breakdowns of these three categories are not readily available, but the southern NAPFSC data for FY 1998 are illustrative. Recall that the southern state-appropriated research budgets were 37 percent of all southern forestry-school and college research funds ($13.3 million of $36.0 million). Total southern forestry-school budgets for instruction were $11.1 million, and for extension $4.6 million. State appropriations dominated the totals for teaching (92 percent) and to a lesser extent extension (72 percent). In total, the 14 reporting southern NAPFSC forestry schools had $26.8 million, or 52 percent of all their funds provided by the states. Total southern shares of funding among the three principal functions were 21 percent of the funds for teaching, 70 percent for research, and 9 percent for extension. If one prorates the southern breakdowns to the national total of $204 million in FY 1997, one would infer that national forestry schools spent about $143 million on forestry research, $43 million on education, and $18 million on extension.

Contributions of the Forest Products Industry

United States forest products firms invest millions of dollars for forestry research. The NAPFSC reports that industry contributed about $14 million in funds to various forestry school programs (NAPFSC, 1999). Industry spending on internal research and development related to forestry—both basic and applied—is more substantial but is concentrated in a few large firms.

Current estimates of forest industry research are based on American Forest & Paper Association (AF&PA) Sustainable Forestry Initiative (SFI) surveys (Cantrell, 2002). AF&PA data indicate that the forest products industry spent $79.5 million on forestry research in 2000 (Table 3-9). Industry research is distributed among five classifications: forest health ($50.6 million), water quality ($7.9 million), fish and wildlife ($8 million), ecosystem management ($7.1 million), and other ($5.9 million). The distribution of research expenditures has an apparent strong focus on environmental issues and problems. The $5.9 million spent on "other" may be the only category strictly devoted to research on enhancing timber productivity.

In addition to the SFI tabulation of forest industry funding, the National Council of Air and Stream Improvement (NCASI) receives separate funds for forestry research.

Table 3-9. Sustainable-Forestry Research Funding by Industry through the Sustainable Forestry Initiative (SFI) Program, in dollars.

Category	1995	1996	1997	1998	1999	2000	2001	1995–2001
Research Funding								
Forest health	30,626,666	37,834,175	42,756,981	43,718,598	44,496,601	50,597,539	35,640,724	285,671,824
Water quality	4,049,731	5,225,464	3,674,481	3,880,018	5,341,227	7,889,960	5,622,303	35,683,184
Fish and wildlife	4,161,855	5,674,969	4,830,664	5,498,191	6,500,639	8,041,715	6,729,270	41,437,303
Ecosystem management	6,504,830	6,566,151	5,059,830	4,621,053	22,751,864	7,090,636	9,506,448	44,813,385
All other research funding	7,679,510	6,687,074	7,553,104	10,620,211	32,539,899	5,857,093	14,724,710	61,870,189
Total research funding	**53,022,592**	**61,987,833**	**63,875,060**	**68,338,071**	**70,551,391**	**79,476,943**	**72,223,455**	**469,475,345**
Research Funding Allocations—Internal vs. External								
Internal research funding	45,523,965	53,591,277	55,165,458	58,186,555	58,579,348	59,266,267	57,987,941	330,312,870
External research funding	7,498,627	8,396,556	8,709,602	10,151,516	11,972,043	20,210,676	14,235,514	66,939,020
Total research funding	**53,022,592**	**61,987,833**	**63,875,060**	**68,338,071**	**70,551,391**	**79,476,943**	**72,223,455**	**469,475,345**

Source: Richard Cantrell, American Forest and Paper Association, Washington, D.C., personal communication, April 2002.

For 1999, NCASI spent about $2.9 million on sustainable forestry ($1.1 million), forested watersheds ($0.6 million), eastern wildlife ($0.4 million), and western wildlife ($0.8 million). Sustainable forestry research included environmental effects of intensive management practices, long-term site productivity, landscape ecology and management, and global climate change. Watershed research included streamside management practices, roads, and cumulative effects. Wildlife research addressed threatened and endangered species management and habitat values of managed forests (Al Lucier, personal communication, 1999).

The forest industry also sponsors research directly through membership in the Institute for Paper Science and Technology (IPST), housed at Georgia Institute of Technology in Atlanta. IPST performs research and transfers technology in paper science, including a program in basic biology and wood properties. Its 1999 budget was about $12 million.

Other Sources of Research Support

Funding of forestry has many other federal sources ("other federal sources"; 18 percent, $37 million; NAPFSC, 1999), including grants from EPA, NASA, NSF, DOE, and the USDA National Research Initiative (NRI) competitive grants program. Non-traditional sources of forestry research funding have increased in recent years. In fact, the $37 million NAPFSC total is $23 million greater than the $14 million reported as coming from (mostly USDA Forest Service) cooperative agreements. The growth of non-traditional sources might be attributed to a leveling off of funding for USDA Forest Service cooperative agreements; an expanding forestry mission for agencies such as DOE, EPA, and NASA; and aggressive pursuit of new sources of funding by university professors.

Many other federal agencies perform forestry research directly, as well as giving external grants. In addition, many schools and departments other than the NAPFSC forestry schools and departments perform forestry research. Separating out all the internal and external forestry research in agencies, and in NAPFSC and non-NAPFSC schools proved to be impossible for this report. Our best estimates of federal agency objectives and funding for forestry research are discussed below.

Funding of forestry-related topics by the USDA's National Research Initiative (NRI) ranged from $4.5 million in 1998 to $9.9 million in 1995 (Table 3-10). That included a fairly stable component of funding for research in use and wood products of about $2 million each year, and a widely fluctuating amount of forestry related research, ranging from a low of $2.2 million in 1996 to the high of $7.8 million in 1995. Funding varies with the merits and success of the individual grants that are submitted to the larger NRI competitive process for all relevant disciplines each year. Total NRI funding ranged from $96 million in 1994 to $88 million in 1998.

The USDA also funded a new program called the Initiative for Future Agriculture and Food Systems (IFAFS) in FY 2000 and FY 2001. This program has components in agricultural genomics, agricultural biotechnology, food safety, new uses for agriculture products, natural resource management, and farm efficiency. In FY 2000, three major grants related directly to forestry were awarded, totaling $8,593,000.

Table 3-10. Federal Funding for Forestry Research by Selected Agency and Program, FY 1994-2000 (thousands of $).

Agency/Program	1994	1995	1996	1997	1998	1999	2000
USDA							
National Research Initiative (NRI)[a]	6,512	9,939	4,121	6,960	4,500	—[b]	—
Forestry	4,244	7,783	2,298	4,424	2,654	—	—
Improved utilization of wood and fiber	2,266	2,156	1,823	2,536	1,846	—	—
Agricultural Research Service (ARS), Initiative for Future Agriculture and Food Systems[c]	—	—	—	—	—	1,924	2,252
NSF[d]							
Division of Environmental Biology	—	5,885	17,906	15,217	17,892	9,409	—
Division of Biological Infrastructure	—	892	1,189	86	1,393	5,171	—
Division of Integrative Biology and Neuroscience	—	0	2,484	1,128	665	730	—
Division of Molecular and Cellular Biosciences	—	0	0	0	0	561	—
Total	—	6,777	21,579	16,431	19,950	15,871	—
DOE[e]							
Terrestrial Carbon Processes Research Program	—	—	—	—	4,934	5,476	4,486
Ecosystems Research Program	—	—	—	—	3,454	3,645	3,133
National Institute for Global Environmental Change Program	—	—	—	—	—	3,722	3,074
Total	—	—	—	—	8,388	12,843	10,693
NASA[f]							
Research and analysis programs					13,100	9,400	13,600
Terrestrial Ecology	—	—	—	—	7,700	6,500	7,800
Land Cover and Land Use Change	—	—	—	—	4,000	2,000	4,300
Earth Observing System Interdisciplinary Science	—	—	—	—	1,000	400	1,100[g]
Natural Hazards (Fire)	—	—	—	—	400	500	200
Forest Topography (Analysis of Radar Data)[h]	—	—	—	—	0	0	200

[a] Source: Cindy Huebner, USDA/NRI, Washington, DC, personal communication, October 1999. Data include projects directly related to forests or forestry. Data exclude indirect forestry-related research (such as, genetics of forest pests and wood products).
[b] Data not available.
[c] Source: Paula Geiger, USDA Office of Budget Program Analysis, Washington, DC, personal communication, March 2000. Data present funding for agroforestry. 2001 president's budget for agroforestry is $2,252,000.
[d] Source: James Edwards, NSF, Arlington, VA, personal communication, December 1999.
[e] Source: Karen L. Carlson, DOE, Germantown, MD, personal communication, March 2000.
[f] Source: Diane Wickland, NASA, Washington, DC, personal communication, February 2000. Data include investments in satellite data analysis specific to forests but not to all vegetation. Data exclude investments in space missions (flight and ground software and hardware) that observe forests.
[g] Does not include new program selections for FY 2000.
[h] This program cuts across the four preceding programs. It was supported as a part of two one-time space shuttle science missions - Shuttle Imaging Radar-C/X-band Synthetic Aperture Radar (SIR-C/X-SAR) and the Shuttle Radar Topography Mission (SRTM).

These included establishment of a tropical forestry center, a sustainable forestry proposal, and a forest biotechnology proposal.

NSF provides grants for research related to forests. The foundation does not have a forestry research division, but many research grants and Long Term Ecological Research (LTER) site projects deal directly with forests, forestry, trees, or wood. Estimates of recent NSF research related to forestry or trees ranged from a high of $21.5 million in 1996 to $15.9 million in 1999. The Divisions of Environmental Biology and Biological Infrastructure provided the majority of this funding.

DOE began an Agenda 2020 research program related to forestry in 1996. In addition, it has funded a variety of forestry-related energy projects for decades. The Oak Ridge National Laboratory is managed for the DOE and conducts direct forest-related research. The previous expenditures by DOE for forestry research were more than $7 million per year. Annual Agenda 2020 expenditures were about $2 to 3 million from 1996 to 1999. Most of those expenditures were targeted toward biotechnology, physiology, soil productivity, remote sensing and wood quality research, but sustainable forestry projects received a substantial share.

EPA has performed or funded a rapidly increasing amount of forestry research, focusing on such issues as global climate change, carbon storage, water quality, and air quality. EPA personnel demurred on providing estimates of their research related to forestry, noting that their work was focused on aquatic resources. They did note, however, that they conduct research on related topics, such as land-use and land cover changes, biogenic emissions from forest canopies and fires, forests as a component of riparian zone restoration, forest fragmentation and habitat, acid deposition and vegetation effects, pesticide effects and exposures to terrestrial vegetation, and whole-watershed assessments. If one uses a somewhat broader definition of forestry-related research, relevant EPA expenditures would be about $10 to 20 million per year.

NASA has funded increasing amounts of research related to forests in recent years. NASA's estimated contribution to forestry research is about $10 million per year, with terrestrial ecology being the largest portion.

There are other sources of government and nongovernment funding of research in forestry subjects, either narrowly or broadly defined. Nongovernment organizations, such as The Wilderness Society and The Nature Conservancy, have applied-research programs that specifically address forestry issues and problems. State forestry organizations such as those of Oregon, Washington, Minnesota, Georgia, and Virginia either have specific funding for forestry research or perform a host of applied studies on ecologic and social issues. Federal agencies—such as the DOI Bureau of Land Management, U.S.Geological Survey Division of Biological Sciences (formerly Fish and Wildlife Service and National Park Service research), and the USDA Natural Resource Conservation Service—perform a wealth of research related to forest flora and fauna. The total amount of their research that is directly related to forestry is not known, but is substantial. In addition, a host of international organizations, ranging from the U.S. Agency for International Development and the World Bank to organizations in other countries, sponsor research related to world forests that provides considerable funding to U.S. and international scientists. In total, those other organizations probably add $10 to 50 million to the more-precise forestry-research funding totals estimated above.

Forestry research could be defined even more broadly—as anything related to the ecology or people associated with the one-third of the nation's total land base classified as forest, or even the world's forest resources. Given a broader definition, the amount of forestry research in the country is indeed very large. However, given that definition, there are many overlaps with other disciplines; it thus provides a blunt tool for assessing the status and deficiencies in our forestry-research capacity. So a narrow enough definition of forestry research is used in our study to estimate trends in investments and accomplishments.

EVALUATING RETURN ON INVESTMENT IN FORESTRY RESEARCH

Investment in forestry research has resulted in diverse benefits, such as lower-cost wood products for consumers, increased income for rural people through improved management and marketing of wood from small woodlots, expanded employment opportunities, improved water quality and flows, maintenance of ecologic integrity and diversity, and enhanced recreation experiences through new recreation-management techniques. Research has led to increased quality and efficiency in the use of all forest resources.

Various studies have examined the returns on investments in forestry research. Bengston (1999) summarized many of the studies that occurred as part of a focused effort in the 1980s; Hyde et al. (1992) published *The Economic Benefits of Forestry Research*; and a few other studies have also been published. Table 3-11 summarizes the results of the studies.

The evaluations indicate that forestry research has consistently had handsome economic rates of return for improvements in individual forest management practice and for wood products research. The average rates of return for wood products research had the greatest returns, ranging from about 15 to 40 percent per year for most conventional research applications. Softwood plywood research had very large returns on research investments, as did wood preservation research, but such breakthroughs are uncommon. The large benefits of forest products research are attributable mostly to the fact that gains are achieved and implemented quickly, and application to a large volume of end products increases net gain. These gains accrue more to wood products producers (large firms) and consumers than to forest landowners or others for whom public research expenditures may be more easily justified.

Timber-management research evaluations also generally found excellent economic rates of return or benefit:cost ratios. Economic rates of return for individual programs such as forest pest management, containerized seedlings, and forest nutrition ranged from 9 percent to more than 100 percent. Benefit:cost ratios ranged from 2.3:1 to 34:1 for fusiform rust research, growth and yield modeling, herbaceous weed control, and tree improvement programs. The one notable exception in these findings was low rates of return (0–7 percent) found for aggregate southern softwood forestry research (Hyde et al., 1992). Hyde et al. (1992) compared aggregate productivity gains for the entire southern forestry sector with aggregate southern forestry research investments. Such aggregate econometric comparisons might provide less robust means of identifying and estimating technical change than individual analyses of production economics and marginal rates of return. Compared with agriculture, aggregate changes in making slight growth improvements in all southern pine production would be expected to be much lower than the spectacular gains or returns one would expect to receive based domesticating wild cereal crops.

Most forestry-research evaluations demonstrate that past gains have been substantial. The fusiform rust research evaluation also estimated the possible incremental gains that could be achieved if fusiform rust were eliminated as a major southern pest. The advent of integrated biotechnology and forest-pathology research makes such a previously unlikely goal possible. Eliminating fusiform rust as a major disease of southern pines could quadruple the calculated benefits of the current tree breeding strategies (Cubbage et al., 2000). Rapid advances in integrated biotechnology, tree breeding, forest nutrition, herbicides, and silviculture have clearly yielded substantial marginal rates of economic return on financial investments (i.e., Yin et al., 1998; Siry et al., 2001) and research investments, although no formal research-evaluation studies have been published.

Forestry research evaluations to date have measured the gains from research that have increased the efficiency of wood utilization and timber management, but they have not captured the gains from productivity sustaining (maintenance) research. An estimated 43 percent of Forest Service research—and probably an equal portion of other forestry research—is aimed at maintaining the existing productivity level, which would decline in the absence of research to deal with disease, pests, and other factors that adversely affect forest productivity (O'Laughlin et al., 1986).

Table 3-11. Return on Investment in Forestry Research.

Research Evaluated	Measures of Economic Impact		
	Marg. ERR[a] %	Avg. ERR[b] %	B/C Ratio[c]
Wood product research			
Structural particleboard (Bengston, 1984)	27–35	19–22	
Lumber and wood products (Bengston, 1985)		34–40	
Timber utilization (Haygreen, et al., 1986)		14–36	
Wood preservation (Brunner & Strauss, 1987)			15:1
Softwood plywood (Seldon & Newman, 1987)	236		
Timber management research			
Forest pest management (Araji, 1981)		60–86	
Tree improvement (Levenson, 1984)			34:1
Forest nutrition (Bare & Loveless, 1985)		9–12	
Growth and yield model (Chang, 1985)			16:1
Containerized seedlings (Westgate, 1986)		37–111	
Herbaceous weed control (Huang & Teeter, 1990)			17-21:1
Timber harvesting (Cubbage et al., 1988)		17	
Southern softwood forestry (Hyde et al., 1992)		0–7	
Fusiform rust (Pye et al.,1997; Cubbage et al., 2000)			2-20:1

[a] Marginal economic rate of return: ERR on additional funds invested.
[b] Average economic rate of return: ERR on total investments; ranges reflect different sets of assumptions.
[c] Benefit:cost ratio, when benefits and costs are discounted back to a common time; ranges reflect different sets of assumptions.

Productivity research is only a portion of public, and perhaps of private, research. Past evaluations of forestry research have not captured the value of economic benefits derived outside the marketplace, such as those related to environmental protection and improvement, and to amenity and recreation values. The prospects for large economic returns to forestry research on nonmarket goods and services also are significant. Research on the nonmarket benefits of the monitoring of wildlife, biodiversity, forest health, and even inventory and analysis also should enhance our management, conservation, and quality of life significantly. One study indicates that the economic benefits of wildland recreation research can be substantial and that society has underinvested in recreation research (Bengston and Xu, 1993). Thus, the rates of return shown in Table 3-11 likely represent conservative estimates of the payoff of public forestry research.

CONCLUSIONS AND RECOMMENDATIONS

Several themes transcend this overview of research capacity. Investment in U.S. forestry research is substantial and more stable in total than commonly believed. But it is fragmented among organizations. Direct USDA Forest Service forestry research personnel and support have declined, and other agencies are increasing their focus on issues related to forestry. Therefore, better information is needed to monitor the status of the inputs to forestry research. Although the Forest Service maintains pertinent information related to much of its research, comprehensive information on forestry research in the United States is lacking.

In 1997, the National Science and Technology Council recommended a framework for integrating the nation's environmental monitoring and research networks and programs, noting that new developments in science and technology provide new opportunities for collecting and organizing data. (National Science and Technology Council, 1997). With current fiscal limitations facing all levels of government, cooperation and efficiency among agencies is essential to the long-term success of individual programs. Following on the need for an integrated environmental and monitoring network, an integrated forestry-research information system is needed for tracking forestry research activities. The initial challenge will be to build on, enhance, and integrate existing databases.

Recommendation 3-1

The Forest Service should enhance its current research-information system and tracking efforts by establishing an improved and integrated interagency system that includes relevant information on forestry research activities, workforce, funding, and accomplishments in all agencies of the U.S. Department of Agriculture, other relevant federal agencies, and associated organizations as appropriate.

Implementation of an enhanced system would require integrating information on forestry research from the Forest Service, agencies in USDA, NSF, DOE, EPA, DOI, and NASA. The system would provide a stronger foundation on which to base decisions for the future. Developing better information on the status of forestry research will require settling on the type of data that should be included in such a system; determining funding and staffing levels of federal, state, university, and nongovernment organizations performing forestry research; noting research priorities; and tracking quantitative and qualitative research accomplishments.

Personnel

Based on the Forest Service survey (2002), 2,186 scientist FTEs were employed at universities, in the Forest Service, or with forest industry in 2001. An estimated total of 1,346 FTEs were dedicated to research, with about 43% at universities, 49% at the Forest Service, and 8% with the private forest industry. About 600 forest scientist FTEs were dedicated to teaching, and 62 to extension. Scientists employed by other federal and state organizations and nongovernment organizations would add perhaps another 50 to 100 to that total.

Whether we have an adequate number of scientists in the requisite disciplines for the future, however, is debatable. Forest Service data support the belief that there have been rapid declines in the numbers of scientists in traditional research areas, such as silviculture, entomology, disease, and forest products. Most other disciplines in the Forest Service experienced declines in the number of scientists employed over the last 15 years. Ecologists have increased in number, but attrition clearly has reduced Forest Service research capacity. Forest Service timber management research probably has declined, but this has been offset by large increases in broad forest management research. Despite perceptions by traditional stakeholders, Forest Service data on funding indicate that environmental research appears to have declined. On the other hand, based on the SFM data tallies by FTE, Forest Service environmental research in biodiversity and ecosystem health research now combines to constitute their largest research area. University research has a broader focus with more emphasis on social science and institutional frameworks. Private industry focuses mostly on productive capacity and soil and water research. Data on disciplines of academic researchers and teachers are not readily available, but experience suggests that academia is unlikely to cover all the shortfalls evidenced by declines in most Forest Service scientific research disciplines.

Recommendation 3-2

The Forest Service should substantially strengthen its research workforce over the next five years to address current and impending shortfalls, specifically recruiting and retaining researchers trained in the disciplines identified as foundation and critical emerging fields of forestry science.

Addressing the rapid decline in scientific manpower will strengthen the Forest Service's ability to respond to short- and long-term research needs. Employing additional full-time permanent researchers, rather than supplementing with temporary employees and post-doctoral students, in fields that are required to address traditional and emerging issues will improve Forest Service continuity and effectiveness in research efforts. Although post-docs and temporary employees are appropriate for some jobs – and do have a place - in many ways they cannot be compared to full-time employees. It is imperative that the Forest Service address the current deficiencies as soon as possible, because the situation is likely to become worse. In the past 8 years alone, the Forest Service has lost over 9000 total employees and during the past 15 years has lost approximately 45% of its scientists. Currently 35% of its workforce is eligible to retire in the next five years and the average age of employees is 55 years, with only five employees under the age of 25 years (personal communication, Mark Rey, USDA). The U.S. Department of Labor substantiates that the number of available workers is decreasing, the average age of the workforce is increasing, the pool of young workers is shrinking, and the number of less educated people in the workforce is increasing (U.S. Department of Labor, 2000). Although employment conditions differ greatly by field and subfield of science (National Research Council 1998), the demand for employees in science and technology in many areas that support important federal missions has outstripped supply (National Science and Technology Council, 2000). The cost associated with strengthening and retaining the Forest Service research workforce is nominal compared with the costs associated with operating under current and projected deficiencies.

Recommendation 3-3

As part of the increase in research personnel capacity and resources, the Forest Service should enhance cooperative relations with forestry schools and colleges.

Partnerships that have evolved between the Federal government and the nation's universities have proven exceptionally productive, successfully promoting discovery of knowledge, stimulating technologic innovations, improving quality of life, educating and training the next generation of scientists and engineers, and contributing to America's prosperity (National Science and Technology Council, 1999). Cooperative research allocations by the Forest Service have decreased markedly from about 15 percent to 9 percent of its budget from 1990 to 1997. The Forest Service should consider designating a larger percentage of its total research budget to the station or research work unit level for extramural research grants that are inter-organizational and cooperative, requiring active involvement, cooperation, and integration of Forest Service, university, and other research partners. The integration of research and education is the hallmark and strength of our research and education system. Two important rationales exist for federal investment in university-based research and these are: (1) the benefits derived from training a new generation of scientists and (2) continuous mutual enrichment that is derived from the relationship (National Science and Technology Council, 1999; National Science Foundation, 1998). The agency could strengthen its relationship with partners if a larger and more openly competitive cooperative grants program existed.

Research Quality, Productivity, and Efficacy

Measuring research quality, productivity, and effectiveness of transferring research to users is difficult. Better oversight and program reviews would help to ensure that organizations are pursuing appropriate strategic directions and implementing them with sound operational programs. The forestry research sector consists of a broad group of public and private organizations. A central organizing body is needed to monitor forestry research and facilitate cooperation among the various organizations. Creation of new federal or state organizations is not necessary, but better oversight and direction from advisory bodies are needed.

Recommendation 3-4

The USDA Forest Research Advisory Committee should focus its efforts in two primary areas: (1) working with USDA research leaders in the Forest Service and other agencies to set research priorities and monitor accomplishments, and (2) coordinating with USDA's Cooperative State Research, Education, and Extension Service and other agencies to help guide research priorities of McIntire-Stennis, Renewable Resources Extension Act, National Research Initiative, and other grant programs.

Those involved in providing focus should include professionals in government agencies, universities, and other relevant organizations as members or ex-officio members. A full-time dedicated professional USDA senior-level director would facilitate operations, serve as communication liaison, monitor forestry research accomplishments, and coordinate site reviews and visits. Those involved would also monitor forestry-research quality and accountability by renewing and expanding the periodic review process, including reviews of McIntire-Stennis projects and Forest Service agency and cooperative agreement research accomplishments. Reasonable intervals for site visits are 10 years for McIntire-Stennis institutions and 5 years for Forest Service research stations.

Advisory groups would help to ensure that research agencies and other organizations are pursuing appropriate strategic directions and implementing them with sound operational programs. Implementing or renewing forestry-research oversight reviews would correspond with the mandates for performance evaluation under the GPRA. Reviews might not necessarily entail additional report preparation, but perhaps more site visits, discussion of research priorities and progress, adaptive management or research programs.

Recommendation 3-5

Universities and state institutions should increase the use of competitive mechanisms for allocating McIntire-Stennis and Renewable Resources Extension Act funds within these institutions, and in doing so, encourage team approaches to solving forestry and natural resource problems as well as integrated research and extension proposals or interinstitutional cooperation.

With goals consistent to the respective Congressional Acts, many universities allocate McIntire-Stennis, Hatch, and Renewable Resources Extension Act (RREA) funding via a merit-based competitive process (for example, see Boxes 3-1 and 3-2). Scientific excellence is promoted when investments are guided by merit review that rewards quality and productivity in research and accommodates for endeavors that might be high-risk but have potential for high gain (National Science and Technology Council, 1999, 2001; National Research Council, 2000).

Clearly, formula-fund allocations are critical for diffusing research throughout the nation, for pursuit of long-term research goals and multidisciplinary research, and for supporting a system in which university faculty appointments are split among some combination of research, extension and teaching. There is a need to preserve the advantages offered by formula funding (University of Idaho, 1983), particularly their facilitation of linked research, extension, and teaching programs (National Research Council, 1996). However, if more competitive approaches were used by universities and state institutions for allocation of formula-based McIntire-Stennis funds, the opportunities for improving the quality and accountability of research funded will be greater. A stronger commitment to addressing the quality and accountability of formula-based research might also provide greater support for funding the critical McIntire-Stennis program at a level closer to that at which it was authorized. The current funding level of McIntire-Stennis is only approximately $21 million, which is less than half its authorized level.

Institutions, or consortia, should concentrate research capital in specific (and perhaps limited) fields of forestry research where they operate best or have some recognized institutional advantage. One of the ways to increase quality and cooperation is to bring federal, state, and private-sector scientists into the academic fabric where needed to augment the expertise of university faculty in preparing future scientists. Collaboration of nonuniversity scientists in the academic fabric could expand the "critical mass" of scientists and educators preparing future scientists.

In addition research oversight and mechanisms, technology transfer should be improved. We have made great strides in many fields of basic and applied research, but resources directed to extension and cooperative efforts have steadily declined. A stronger delivery system must be developed.

Recommendation 3-6

The U.S. Department of Agriculture, together with universities, should develop means to more effectively communicate existing and new knowledge to users, managers, and planners in forestry.

If we are to achieve broadly recognized forestry research and development goals, our technology transfer and extension capability should be enhanced. Almost 10 million nonindustrial private landowners rely on extension, communication, and transfer of research results to make informed decisions (National Research Council, 1998). Universities, government, and private organizations should work together to improve mechanisms for communicating research and technology.

Fiscal Strength

At least $400 million is spent on forestry research each year by the various research organizations in the United States, and the total might well exceed $500 million. Funding includes about $200 million for Forest Service research and $204 million for research in professional forestry schools, colleges, and departments. NAPFSC data indicate that forestry schools received about $23 million of their external research funds from non-Forest Service grants and $12 million from Forest Service cooperative agreements in 1998. The USDA has provided other funding through NRI and IFAFS, in the amount of approximately $10 million per year. Including the data reported in the SFI and NCASI research, the forest industry spends at least $70 million per year in forestry research and probably far more on wood and paper research. State agencies spend a few million dollars per year in total on applied forestry research. Federal agencies other than the Forest Service were unable to provide definitive estimates of their funding of forestry research, but DOE, EPA, NASA, DOI, and NSF spend at least $10 million per year on research specifically related to forests. Total forest research expenditures in the United States were about $530 million in 1998.

Trends in university and nonfederal forestry research are difficult to assess. Non-Forest Service federal, state, and nongovernmental organization forestry research has increased in recent years despite fairly static funding in Forest Service research funds. Forest industry research also appears to have increased in the last 5 years, although it is concentrated in a few firms.

Toward Greater Capacity

The overview presented here suggests that financial and human investments in forestry research, construed narrowly, are substantial and that return on investment is high. Forestry research may be defined more broadly to include much of natural resources research. In either case, the nation has moderate capacity to discover new knowledge about forest resources. However, the nation's forestry-research capacity and investment in research, particularly in Forest Service research, have declined sharply in the last decade. Many scientific disciplines appear to have dwindling numbers of research scientists and dwindling expertise despite rapid increases in pressing problems regarding the productivity, health, management, and protection of our nation's forests. Those trends are important and must be addressed without delay, given the rapidly increasing number of challenges and issues facing forests and forestry research.

4

Preparing Forestry Scientists and Users of Forestry Science

Forests must accommodate a wide variety of uses and benefits. As a result, foresters are asked to protect and enhance biologic diversity, protect water quality in well-managed watersheds, provide habitat for game and nongame species, create recreational opportunities ranging from wilderness environments to developed campgrounds, and provide wood and non-commodity forest products. Long-standing issues, such as how to provide forest products in an economically efficient and environmentally sound manner, and many new issues—such as environmental justice, inequities in resource availability, habitat fragmentation, endangered species, and urbanization, have entered the public discourse. Both new and old issues often seem acutely complex, and their solutions are rarely straightforward.

The boundaries that define forestry are expanding rapidly to accommodate the vast array of benefits and values associated with the forest. Population viability analysis, ecologic services, landscape management, cumulative impacts, amenity-based development, recreation carrying capacity, adaptive management, conflict resolution, collaborative learning, and other subjects are being proposed as basic curriculum elements. Traditional sustained-yield approaches that focus on commodity production are giving way to comprehensive and integrated approaches that emphasize ecologic and social sustainability. The new approaches focus on maintaining and restoring ecosystem integrity and long-term productivity while guiding appropriate human uses of natural resources.

THE FUTURE OF FORESTRY EDUCATION

Given the expanding view of forest management, how do we educate the next generation of foresters? To be successful managers, forestry graduates must be broadly educated and possess a variety of skills, tools, and technologies in order to understand the ecologic and social processes affecting ecosystems (Sample et al., 1999; Bentley, 1999). Graduates must also have solid skills and fundamental knowledge in basic sciences. In response to comprehensive and integrated approaches to resource management, the challenge is to find the means by which focused education, interdisciplinary systems thinking, and communication skills can be developed and applied by forestry professionals.

The need for a focused education and interdisciplinary thinking might appear to be contradictory. Yet the challenge for academic institutions in educating the next generation of resource managers is to provide each student with a common set of skills that include oral and written communication, interpersonal skills, problem-solving, and critical thinking and with the ability to implement the skills in natural-resource management. In addition to the skills, there is a need for basic knowledge in a discipline that can be applied in a holistic context.

It has been recognized that there is a need to promote and achieve disciplinary integration and apply the resulting knowledge to complex social and biologic problems (see, for example, the article "The Employer's Perspective on New Hires" in the September 1999 issue of the *Journal of Forestry*). Progress toward those goals has been thwarted by discipline-focused faculty, a tradition of reductionism in conducting research, and fragmented curricula in many academic institutions. But integration can come from achieving both depth and breadth in an academic program that includes both teaching and research. One set of courses can ensure depth in a discipline, and another set can promote breadth of exposure and connections to other disciplines.

Creative approaches to disciplinary integration at the undergraduate level have been implemented and evaluated in several forestry programs, including those of the University of Vermont (Ginger et al., 1999) and Northern Arizona University (Fox et al., 1996). Those experiments in teaching and learning generally involve the study of natural resource issues and use core and capstone courses that blend the biologic and social sciences. Without question, integrating across disciplinary boundaries places additional burdens on instructors. Practical matters need to be resolved, such as defining content, coordinating schedules, and establishing teaching assignments (Ginger et al., 1999). More important, there needs to be an intellectual commitment to operating outside the comfort of one's discipline. Obviously, administrative support is critical to success (Fox et al., 1996). Finally, there are pedagogical issues. For example, social scientists are more apt to use discussion, debate, case studies, and team efforts in teaching, whereas biologic scientists have a tradition of using lectures to transfer information in the classroom (Ginger et al., 1999).

Many of the prerequisites for integrated teaching and learning also apply to integrated research. It seems logical to operate as a member of a diverse team of scientists to address multifaceted problems, but true integration is rarely achieved.

Reductionism remains the more common research approach, and most forest scientists believe that the greatest gains in knowledge come from a strong disciplinary focus. According to Ross Whaley, our failure to integrate might not be so much an inability to conduct interdisciplinary research as an inability to integrate and synthesize the results of our research (unpublished presentation provided during the National Research Council's workshop on National Capacity in Forestry Research). In other words, it is more a *thinking* problem than a *doing* problem (unpublished paper presented to the National Research Council's Workshop on National Capacity in Forestry Research, July 15, 1999, Washington, DC). As Whaley recognized, the "ability to integrate vast amounts of information from many disciplines and a broad array of viewpoints has not been adequately honed in their formal education or their apprenticeship. This is not easy. It takes a special kind of formal education that must be refined through experience."

Successful implementation of a broader, more integrative approach to resource management necessitates higher levels of interaction between researchers and managers than has been the norm. The relevance of research information to resource problems and environmental issues needs to be clearly identified and communicated. Research information should continue to be subjected to peer review to ensure the quality of the research enterprise, but it should also be communicated in forms that are useful to resource managers, planners, and policymakers once it has passed the test of peer review. To judge from public reactions to land management, there is much room for improvement in communicating science to a broad audience.

The successful management of any complex system—biologic, social, or physical—requires first the knowledge of the fundamental concepts and laws that govern the operation of the system. Too much attention has been paid in forest science to the collection of data and facts; too little effort has been invested in developing the theoretic framework for the social and biologic sciences that are commonly applied to resource management. Without a framework, we are limited to an endless litany of empirical studies whose results have predictive value for a narrow range of conditions. Without an organizing structure or theory, information is merely a collection of observations and unrelated fragments of data. Without structure, it is difficult to learn from experience or to extrapolate into the future. A continuing challenge facing forest scientists (and those in other parts of the biologic and social sciences) is to develop an explanatory and predictive system of concepts, theories, and laws. Meeting that challenge is a major issue for forestry education.

Discussions about the need for a focused education in natural-resources management and for interdisciplinary systems thinking generally center on how to achieve an appropriate balance between breadth and depth in the curricula. There seem to be two competing needs—more opportunities to build interdisciplinary perspectives and a need to provide greater depth in some fields. Breadth is needed at both the knowledge level (e.g., a student of forest management understands something about social systems) and the process level (e.g., students understand problem-solving approaches). A challenge is to design opportunities for breadth-building without diminishing the capacity for building the depth needed for students to become successful professionals.

There are practical and sometimes legal limits to the number of credit-hours that can be required for a baccalaureate degree. Each course added to the curriculum should be evaluated relative to its contribution to the faculty's vision of an education necessary for a professional forester and to the stated mission of the school. The primary need is to identify a general educational and professional core of courses essential for student development and then supplement the courses with transdisciplinary, quantitative, and holistic educational experiences. The increased breadth of forestry education does not need to fragment forestry curricula.

TRENDS IN ENROLLMENT AND GRADUATION

Trends in forestry enrollment and degrees awarded provide empirical evidence about the implicit interests in professional employment and research needs. Table 4-1 summarizes data on enrollment in and degrees awarded in forest and wood science programs from 1989 to 1998 (FAEIS 1999a, 1999b). Student enrollment and the number of degrees awarded at all levels have increased throughout the 1990s. Much of the enrollment gain was achieved by 1992-1993 for all degrees and peaked at that time for graduate enrollment, which has since stabilized. The number of degrees of all levels awarded has generally increased continually throughout the period, indicating an improving completion rate.

Baccalaureate enrollment in forest-science programs increased by 53 percent from 1989 to 1998; degrees granted increased by 61 percent. Master's and doctoral enrollments and degrees granted increased less during that period. Master's enrollment increased by 12 percent and degrees granted by 27 percent; doctoral enrollment increased by 5 percent and degrees granted by 30 percent. As of fall 1989, baccalaureate enrollment accounted for 73 percent of all the forest science students; by fall 1998, it accounted for 79 percent. Baccalaureate degrees granted also increased their share of total completions. In fall 1989, bachelor's degrees accounted for 68 percent of all forest science degrees granted; by fall 1998, they accounted for 73 percent.

Many forestry programs now include education and research beyond the traditional "forest science." The Food and Agriculture Education Information System (FAEIS) collects similar data on natural resources, agricultural sciences, and other programs. Those data provide another perspective on the trends in natural resources education toward college degrees. Table 4-2 summarizes enrollment data by major program and degree level from fall 1993 to fall 1999 (FAEIS 1999a).

From fall 1993 to fall 1999, the number of students in natural resources programs was fairly stable. Doctoral programs realized a slight increase in enrollment, but other degrees had fewer students enrolled. Agricultural science programs realized marked increases in undergraduate enrollments in fall 1996 and then declined. Graduate enrollment in agricultural programs fluctuated but generally declined. Forest science accounted for about 8-12 percent of the reported enrollment in fall 1999.

Table 4-1. Enrollment and Degrees Awarded in Forest Science Programs, 1989-1998.

	Number of Students									
Degree	1989	1990	1991	1992	1993	1994	1995	1996	1997	1998
Enrollment										
Bachelor's	4830	5416	5585	5983	6890	6839	7007	7660	7917	7369
Master's	1103	1058	1167	1365	1341	1267	1137	1248	1238	1236
Doctoral	715	693	714	722	792	741	674	744	720	749
Awarded										
Bachelor's	956	929	850	1114	1116	1239	1242	1431	1431	1536
Master's	336	288	330	368	379	384	361	400	384	427
Doctoral	110	107	104	113	109	117	126	122	126	143

Source: Food and Agriculture Education Information System (FAEIS 1999a, 1999b).

Table 4-2. Enrollment in Forestry, Natural Resources, and Agriculture Programs by Program and Degree Level (NAPFSC/SAF Schools), 1993-1999.

	Number of Students						
Program	1993	1994	1995	1996	1997	1998	1999 (% of total)
Bachelor's:							
Forest sciences	6890	6839	7007	7660	7917	7369	6650 (8.4)
Natural resources	17407	16279	16815	17692	17209	16370	15634 (19.7)
Agricultural science	42464	43619	46415	78209	54264	51565	51352 (64.5)
Other	4840	5067	5744	5823	6610	5939	5922 (7.4)
Total	71601	71804	75981	109384	86000	81243	79558 (100.0)
Master's:							
Forest sciences	1341	1267	1137	1248	1238	1236	1162 (12.0)
Natural resources	2715	2352	2557	2566	2481	2376	2339 (24.1)
Agricultural science	6295	6190	6210	6004	6527	5754	5802 (59.9)
Other	509	403	464	450	410	421	389 (4.0)
Total	10860	10212	10368	10268	10656	9787	9692 (100.0)
Doctoral:							
Forest sciences	792	741	674	744	720	749	751 (9.3)
Natural Resources	1068	1008	1079	1168	1008	1133	1205 (15.0)
Agricultural science	6610	6501	6176	6063	6357	5897	5957 (74.0)
Other	95	79	128	136	139	121	134 (1.7)
Total	8565	8329	8057	8111	8224	7900	8047 (100.0)

Source: Food and Agriculture Education Information System (1999a).

Natural resources programs accounted for a larger share of the undergraduate and master's enrollment (20 percent and 24 percent, respectively) but only 15 percent of Ph.D. enrollment.

FORESTRY AS AN ACADEMIC SUBJECT

As reflected in the trends just described, forestry as an academic subject has evolved along two paths: undergraduate and graduate education. They are not mutually exclusive, but their evolutionary history is important for answering the question, What forestry curricula will prepare foresters to conduct and use research effectively in the coming century?

The study of forests (in the sense of scientific research and academic scholarship) is open to all disciplines, and most have had some influence in what we know about forests. In a more focused way, specialized researchers in many disciplines participate directly in forestry-research organizations. Many people who do not have a professional forestry degree (or another resource-management professional degree, such as in wildlife biology or landscape architecture) contribute directly to solving forest and forestry problems. Their recruitment to forestry is only a matter of curriculum insofar as exposure to forest and forestry research challenges during their education, particularly their post-graduate education, can attract them to forest subjects and institutions.

On the following pages, we examine the two major paths—undergraduate and graduate—of professional forestry education, discuss their major curricular trends, and try to match the trends with a vision of future research needs.

CURRICULUM AS A CONCEPT

The idea that the quality of an education is determined largely by careful faculty specification of subjects to be studied is old but not universal. In some important ways, the notion of curriculum is antithetical to the notion of a "liberal education". In the purest form of the latter, a student's curiosity confronts an array of subjects, teaching styles, and possible degrees of specialization within broad subjects. Each student selects from that array to create a "curriculum" unique to his or her goals. In an important sense, this is the best curriculum possible if eagerness to acquire knowledge, freedom of inquiry, and development of the individual intellect are important educational values. Individual choice is approximated, to a greater or lesser degree, by most undergraduate liberal arts programs, in which breadth of knowledge, the ability to think independently, and intellectual maturity are important goals.

In both undergraduate and graduate professional programs, however, requirements for specific "professional skills", the accreditation of professional-degree programs by professional societies, and in some cases external licensing requirements force (or are thought to force) a tighter external specification of the content of a course of study. The degree to which those external forces are actually important and the subjects and teaching methods that should be used to respond to them have been subjected to intense debate among forestry educators and professionals in the United States for a century, and longer in some other places. The story continues to unfold, but trends are apparent.

In examining trends, it is important to remember that the central notion of curriculum is that if the specified subjects in the specified amounts are learned successfully, an effective professional education has been offered and received. A given curriculum is assumed to be a set of input specifications, like a recipe for a stew. Unlike the stew, however, the product is probably as much a result of other factors, inside and outside the formal educational experience, as it is of the curriculum. Many think, for example, that the intrinsic capability of the student is the major determining factor in professional success (in Iowa it is called the "Grinnell effect"; that is, if you let only smart ones in, you usually let only smart ones out). Others believe that exposure to people and situations that inspire students and cause them to think is more important for a high-quality professional education than curricular specifications. Still others believe that the emphasis should be on the overall quality and subject mix of the whole faculty; if students are allowed to choose from an array of subjects and teachers that are all important, relevant, and professionally useful, there is no need for further specification. These important disagreements probably mean that there will never be a single, "received" forestry curriculum. Given the diversity of forests and people's views and values related to them, that is probably good.

MODELS FOR FORESTRY EDUCATION

As forestry education began in earnest in the United States in the late 19th and early 20th centuries, two models were applied virtually from the outset. One regarded forestry as a graduate professional subject similar to law and medicine, the other regarded forestry as an undergraduate pursuit more akin to the existing models for engineering and agriculture. The second, not surprisingly, took root most vigorously in the land-grant colleges, which by the beginning of the 20th century had substantial experience with engineering and agricultural curricula.

Current enrollment data indicate that the land-grant-college professional-school model has become dominant for both undergraduate and graduate programs. Land-grant colleges and other state-assisted forestry schools now educate all the undergraduate foresters in programs accredited by the Society of American Foresters, SAF (7419 in fall 1998). They also enroll 96 percent of the master's students and 99 percent of the doctoral students in the forest sciences (FAEIS 1999a). In fact, Yale University was the only private university even reporting any graduate students in forest sciences in the FAEIS system—at 50 master's and 8 doctoral students.

Yale was the original home of the graduate model, and later it emerged at other non-land-grant institutions, such as Harvard, Duke, and the University of Michigan. Yale, Harvard, Duke, and Michigan, relatively early in the evolution of U.S. forestry, began to offer professional doctorates (the doctor of forestry, or DF, degree). Despite the small proportion of the enrollment at these private or graduate-only educational institutions, these institutions continue to serve as opinion leaders on directions for forest science. Indeed, their expansion beyond narrow forestry curricula was emulated by, or at least occurred in parallel with, that at many other institutions.

The evolution of both undergraduate and graduate models has been similar in many respects. From an early emphasis on the biologic and physical aspects of forests, the social sciences (first economics and then the others) have slowly found a place in the curriculum. Forestry departments and schools were ambivalent about the emergence of modern ecology for a long time. Silviculture and ecology, like silviculture and economics, were for a long time, and to some degree still are, uneasy academic partners.

There has been a similar broad agreement between the models on what subjects were important to study. The initial emphasis on trees and wood as the major components of forests, from both biologic and economic points of view, has continually been modified by increased emphasis on other forest components and disciplines related to them. Both models accepted relatively uncritically a utilitarian view of forests; that is, forests are important because of what they can do for people.

But the differences between the models are profound and will probably determine the future of forestry education. The "graduate professional" model has as a basic tenet that a wide variety of undergraduate programs are suitable as a starting point for a forestry education but also that an undergraduate course of study is necessary before beginning a forestry education. The "undergraduate professional" model says that a forestry professional can be created through four years or more of relatively highly specified study at the undergraduate level strengthened with basic liberal arts components meeting university core education requirements and that further formal study, although probably beneficial, is not necessary. The undergraduate model has much to recommend it. It is less expensive in time and money for the student. There are benefits to society at large: the United States has many forests, and large numbers of foresters are needed to manage them; the efficiency of the undergraduate curriculum; and the vast capacity of (particularly) land-grant universities in supplying the numbers needed. The graduate model, in contrast, offers more liberal breadth and scientific depth. It requires a higher caliber student, as demonstrated by the requirements for excellent undergraduate degree grade point averages and good Graduate Record Examination scores to enter a graduate degree program.

Numerically, undergraduate programs dominate, as indicated in Table 4-1. In fact, their share of professional forest science enrollment has actually increased over the last decade. The focus of the undergraduate forestry programs has probably diverged from traditional land management that was typical of forestry programs. All retain the core biology, measurement, management, and policy courses, as required by the SAF (1998) accreditation procedures. But more offer specialization, such as in business, forestry operations, urban forestry, or environmental science (e.g., Bentley, 1999). A greater share of graduates with B.S. degrees are obtaining employment in wood procurement and forestry consulting in the South, and environmental consulting and planning elsewhere. Employment in land management positions has actually declined at the B.S. level (Cubbage et al., 1999). These trends are probably duplicated at the graduate level.

Broad Trends in Forestry Education

Despite the relative increase in undergraduate professional education and its increasing bifurcation along production or environment lines, graduate education remains a crucial component of educating the next generation of resource managers and forest scientists. The importance of natural-resources and environmental programs is obvious, in that they have almost twice the enrollment of forest science programs. Many distinguished former forestry schools or colleges have become natural-resources or environmental-science colleges. In many places, the numerically dominant course of study in "forestry" schools is no longer forestry, but rather a broader curriculum called "natural resources" or "environmental studies". The number of these programs has grown considerably in forestry schools and in other units of universities, such as colleges of arts and sciences. In some cases, they undoubtedly compete for students that formerly would have enrolled in forestry and other professional curricula.

Increasingly, predictions of or calls for a graduate degree as the first professional degree are heard from academic and professional sources (Gordon 1984, Wallinger 1991). The parallel with other graduate professions, such as law, medicine, and business is increasingly drawn. A graduate degree has become almost necessary to advance to higher positions in many organizations that employ foresters. It appears that the broad and broadening knowledge that forestry requires is leading forestry education to a model in which broad undergraduate education, both liberal and professional, is followed by a graduate education that combines elements of science, business, and traditional forestry subjects. This graduate education often leads to a master's degree (master of forestry or equivalent), but there are calls for it to be a doctoral degree (Wallinger, 1991). Whether such predictions will eventually be borne out by educational practice is unknown, but many other issues and trends in forestry education remain important.

Research capacity and education at the graduate level are still closely related to trends and needs in the undergraduate forestry curriculum. At the very least, undergraduate enrollment and teaching appointments tend to drive the nature of the faculty appointed with state funds. Many of the students who enter forestry graduate programs were forestry undergraduate students. Thus, undergraduate education should be as broad as necessary to cover forest science well and as deep as possible to provide insights about basic principles and skills. While curricula continue to broaden from "forestry" to "natural resources," traditional disciplines (e.g., botany, zoology, physics, etc.) and departments have an essential place at the undergraduate and graduate levels in educating prospective forest scientists.

An ad hoc SAF committee (SAF 2000) considered the relevance of SAF accreditation and the form it should take. The merits of a broad general education, balance and depth in professional requirements, and the need to provide instruction in more disciplines while states are reducing the number of credit hours required for graduation are among the issues faced by forestry programs. Professional forestry courses at the graduate level are particularly important for research degrees. Those and many other issues could engender a book by themselves, and they reflect the larger debates about what constitutes forestry and how it should be taught. We will simply draw on the preceding discussion to make recommendations about research implications.

WHAT ABOUT RESEARCH?

What should a research organization—such as the research branch of the USDA Forest Service, the largest forestry-research organization in the United States—think and do about the broad trends described above? There are several very positive potential outcomes for research organizations:

- Undergraduate students should be provided with a broad education in the traditional fields of forest science, but opportunities for specialization or diversification should be encouraged for later graduate education. Similarly, regional differences and employment needs should be recognized.
- Insofar as graduate education produces greater exposure to research (and it usually does) research organizations will benefit in two ways: practitioners or managers will understand better and be more receptive to research results and research cultural values, and more people will be attracted to careers in research.
- One characteristic of forestry schools has been a reluctance to teach broadly in their university because of the needs of their own undergraduate majors. If less emphasis is put on undergraduate majors, teaching capacity might be freed to teach broadly, and this in turn might result in the attraction of more people and a greater array of disciplines to forestry and forestry research.
- As forestry becomes more complex, so does forestry research. The greater breadth of experience of the graduate student should be helpful in confronting this complexity creatively.

There also are issues of concern:

- Will a greater emphasis on breadth and integration decrease the supply of highly specialized researchers? Will fewer entomologists or molecular biologists be available to forestry?
- Many—including many in government and private forestry organizations—think that adaptive management (i.e., an integrated, multidisciplinary approach for confronting uncertainty in natural resources issues) will greatly increase in importance, but there is little evidence that this is an effectively taught curriculum element. How can this topic be effectively included in forestry education?
- Will an undergraduate trend toward broader curricula lead many of the ablest students away from forestry organizations?
- Will forestry schools provide the incentives and environment to produce sufficient gender and racial diversity for forestry-research organizations in the next century?

- Is basic science adequately represented in forestry schools and curricula in an increasingly results-focused era and society? Will the foresters of tomorrow be able to understand the value of basic knowledge, or will they regard its production as someone else's business?

Are important disciplines being lost because of "market" trends? Questions about forest protection (such as the results of long suppression of fire in the West and the growing importance of "invasive alien species", such as the longhorn beetle) seem to be increasing, but the supplies of entomologists, pathologists, and fire scientists seem to be stable or decreasing.

WHAT ABOUT CURRICULA?

No matter which broad educational model is followed, some elements must be included in educational programs. To achieve a balance between depth and breadth and to meet the challenge of producing scientists and those who can effectively use science, the intellectual goals for educating forestry students in both content and process should include many of the following elements:

- Mastery of research methods (problem definition, research design, analytic tools, problem solving) in areas of interest to the student
- Sufficient breadth of knowledge and skills necessary for working with diverse groups both within and outside the student's field of study
- Competency in communicating with diverse audiences
- Specialized knowledge that provides an in-depth understanding of concepts, processes, and interactions within a scientific discipline
- Integrative thinking that promotes a broader understanding about the application of specialized knowledge

Although specialization is unavoidable, indeed desirable, education should still lead to the capacity for broadly informed judgment, and this capacity requires an education that is both broad and basic. By its very nature, natural-resources management is a multi-disciplinary subject requiring the integration of the biologic and social sciences. A goal common to schools of natural resources is to generate knowledge through research and teaching and to help to apply it to meet the full range of human needs on a sustainable basis. That goal is best accomplished through joining disciplines and approaches. Integrating fields of knowledge in natural-resources curricula remains an important challenge for most schools of forestry and natural resources.

One area that might be considered is the preparation of professionals to transmit scientific information to managerial audiences. In land-grant colleges, these professionals would be extension staff. Such programs as the master's program at Oregon State University that specifically targets extension and other educational outreach are few, but they provide an important curricular path if forestry science is to inform management effectively.

With those ideas and trends as background, one might ask, How are we doing in developing capacity for research in forestry? Are we producing the diverse cadre of scientists necessary to meet the needs of a changing forestry profession?

ADEQUACY AND CAPACITY OF UNIVERSITY PROGRAMS TO MEET NEAR-FUTURE NEEDS

We know that university programs are the primary source of new forestry scientists. Doctoral programs produce scientists that become employed in universities, government research organizations, and the private sector. Do they have the capacity to supply new scientists to meet near-future needs for forestry-research organizations? Are they likely to be able to graduate scientists that will enhance the gender, racial, and ethnic diversity of research organizations?

Disciplinary Breadth of Forestry Education

A huge number of disciplines are involved in forestry, and scientific capability is needed in all of them. Over the last couple of decades, many new fields of inquiry have emerged and some traditional fields have declined in importance. The forestry and natural-resources schools have recognized the changes, and the mix of faculty expertise has changed. Forest and range ecology, recreation and tourism, remote sensing and spatial analysis, and natural-resources social sciences have all experienced substantial increases in capability. Forest entomology and pathology have undergone change to focus on integrated forest protection, and forest genetics has become part of the forest-biotechnology arena.

In sum, it is difficult to identify any field that is not represented in the capability of forestry and natural resources schools collectively, but it is clear that some schools emphasize some programs, and other schools emphasize others. Some fields that are represented in only a few schools are fire ecology and behavior, pulp and paper science, wilderness management, forest soils, tropical forestry, forest biotechnology, forest products marketing, and forest engineering and harvesting. New scientists can be produced in those fields, but there are few faculty members to produce them.

In addition to forestry and natural-resources schools around the country, scientists can be produced by many other programs such as botany, biology, entomology, pathology, soils, economics, sociology, and political science. However, those programs usually have a broader focus than forestry and natural resources, so there might be only one or two faculty in them on a campus who are interested in forestry and natural resources; their contribution to the development of forestry and natural-resource scientists would be small. Building bridges with those other programs to increase the probability of developing forest scientists is important.

Tables 4-1 and 4-2 summarized enrollment and graduation trends for forestry programs and enrollment for broader natural-resources and agriculture programs. Table 4-3 summarizes university forest science program enrollment from fall 1993 to fall 1999 (FAEIS, 1999a). These data should help to inform discussions of university capacity.

The forest science categories are fairly broad, but some observations seem worthwhile. At the doctoral level, probably the most interesting finding is the relative stability in the number of students enrolled over the last five years in each of the 15 identified categories. General forestry had the largest enrollments, about 119 to 183 students nationally; forest management had the second largest enrollment, at 141 to 183; forest biology had the third largest enrollment with 128 to 153 ; forest sciences had 79 to 145; forest mensuration, 17 to 23; urban forestry, 2 to 5; and wood science, 53 to 71. Most other disciplines had fewer than 10 doctoral students. The variation in the number of master's students was greater, with sustained increases in the number of students in forest-products technology, forest biology, and urban forestry. Sustained enrollment decreases occurred in forest engineering and forest management.

The forestry and natural-resource schools are quite limited in their capacity for production of forestry doctorates beyond these levels. The FAEIS statistics for fall 1999 indicate that only 12 forestry graduate programs had 20 or more doctoral students enrolled, and the students in these 12 schools made up 69 percent of the 764 Ph.D. students enrolled in forest sciences nationally. The 12 schools are well distributed geographically with three in the Northeast, three in the North Central U.S., two in the South, and four in the West (FAEIS, 2000).

If one considers the forestry and natural-resources doctoral programs combined, over twice as many (29) schools enrolled at least 30 forestry/natural resource Ph.D. students in fall 1999. The students in these 29 schools comprise 90 percent of the 2256 forestry and natural resource Ph.D. students enrolled nationally. Six of these schools are in the Northeastern, six in the North Central, eight in the Southern, and nine in the Western regions (FAEIS, 2000).

Numbers of Scientists

Ability to produce doctorates in the necessary fields depends on several factors, including the disciplinary and integrative orientation of faculty; the stage of faculty in their own careers; the supporting programs, course work, and faculty at individual universities; the numbers of qualified students interested in particular fields, and the assistantship and research support available (Box 4-1). Some students who are conducting large research projects as part of their degree program are guaranteed complete support though their advisors while other students may have guaranteed support for tuition and stipend, but must independently seek support for their research. The latter is typically true if the student is conducting research that is outside their advisor's research program. With the broadening of the field of forestry, there has been a growing number of fellowships available to students in forestry. However, there is also a growing

Table 4-3. Enrollment in Forest Science Programs by Academic Specialization (NAPFSC/SAF Schools), 1993-1999.

Specialization and Degree	Number of Students						
	1993	1994	1995	1996	1997	1998	1999
Forestry, General							
Bachelor's	3098	2956	3086	3467	3349	2882	2462
Master's	360	322	235	285	342	379	354
Doctoral	183	157	119	163	161	181	170
Forest Harvesting and Production							
Bachelor's	0	5	8	76	94	70	50
Master's	6	0	0	4	5	5	3
Doctoral	3	0	4	5	5	3	4
Forest Products Technology							
Bachelor's	55	60	67	116	155	115	147
Master's	20	6	8	14	21	18	17
Doctoral	53	9	3	6	17	6	4
Timber Harvesting							
Bachelor's	7	8	9	13	4	0	0
Master's	3	0	0	7	4	2	0
Doctoral	1	0	2	0	0	0	0
Forest Sciences							
Bachelor's	377	396	424	421	452	456	433
Master's	197	201	208	207	182	167	206
Doctoral	87	95	90	79	90	109	145
Forest Biology							
Bachelor's	325	359	396	456	461	477	464
Master's	152	155	197	228	225	201	174
Doctoral	128	139	142	151	153	143	128
Forest Engineering							
Bachelor's	188	198	194	218	264	292	216
Master's	72	21	14	11	8	8	15
Doctoral	37	16	11	8	8	7	10
Forest Hydrology							
Bachelor's	0	0	10	15	22	20	18

Table 4-3. Enrollment in Forest Science Programs by Academic Specialization (NAPFSC/SAF Schools), 1993-1999. (continued)

Specialization and Degree	Number of Students						
	1993	1994	1995	1996	1997	1998	1999
Master's	47	32	43	30	24	35	27
Doctoral	11	11	12	7	6	9	5
Forest Management							
Bachelor's	1612	1554	1529	1577	1778	1783	1752
Master's	286	287	264	282	258	206	185
Doctoral	141	147	161	183	146	147	151
Forest Mensuration							
Bachelor's	0	0	9	2	3	0	0
Master's	39	22	13	13	19	26	32
Doctoral	21	17	23	21	17	18	20
Urban Forestry							
Bachelor's	88	111	123	124	129	169	169
Master's	13	24	28	27	23	32	30
Doctoral	5	4	4	2	5	3	4
Wood Science							
Bachelor's	278	232	260	351	335	334	281
Master's	71	64	70	85	70	76	64
Doctoral	71	58	53	56	71	58	53
Pulp and Paper Technology							
Bachelor's	607	634	585	606	592	581	543
Master's	10	10	9	11	14	15	11
Doctoral	7	8	12	18	7	12	9
Forest Soils							
Bachelor's	0	0	2	8	0	4	5
Master's	11	6	4	8	14	12	15
Doctoral	3	0	3	5	4	3	4
Forest Sciences, Other							
Bachelor's	54	104	97	116	132	158	110
Master's	20	21	23	36	37	38	29
Doctoral	32	36	28	32	40	42	44

Source: Food and Agriculture Education Information System (FAEIS 1999a).

number of students competing for the grants. These include the Environmental Protection Agency (EPA) STAR (Science to Achieve Results) Fellowships, National Aeronautics and Space Administration Global Climate Change Fellowships, and the Morris K. Udall Scholarship and Excellence in National Environmental Policy Foundation Fellowships, among others. Many graduate students spend a great deal of time securing funding for research.

The statistics show that the production of doctorates is fairly stable, but the distribution over fields is uneven. For example, over the three-year period 1996 to 1998, the production of doctorates nationally was 206, 232, and 248 in natural-resource fields and 130, 116, and 143 in the forest sciences (FAEIS; 1997b, 1998b, and 1999b, 2000 respectively). In the natural-resource fields, environmental studies and sciences, wildlife, and renewable natural resources are consistently the fields with highest production of doctorates. In the forest sciences group, general forestry, forest management, and forest biology consistently have the most graduates. Timber harvesting, forest harvesting and production, forest engineering, forest hydrology, forest soils, forest mensuration, and urban forestry have had few or no graduates. Some graduates with expertise in the fields in which there were few or no recorded graduates probably are in the general forestry and forest management categories (forest soils and mensuration might be good examples), but the numbers of such graduates are probably small.

The number of doctoral students in the nation's forestry and natural resource schools generally mirrors the graduation statistics. For the 1996, 1997, 1998 and 1999 academic years the fall enrollments were: 1434, 1289, 1390, and 1492 doctoral students in natural resource programs, respectively, and 736, 730, 741, and 764 in forest science programs, respectively (FAEIS, 2000). The natural-resource fields with the largest numbers of students were wildlife and environmental studies and sciences. The forest science categories with the largest numbers of students were general forestry and forest management. The categories with consistently few or no students were timber harvesting, forest harvesting and production, urban forestry, and forest soils.

Box 4-1 Graduate Student Support

Graduate student support varies across institutions (National Research Council, 1995). There are a variety of mechanisms that are used by graduate students to finance their education. Many students pay for programs with financial aid, such as federal loans. Other sources of funding include assistantships, such as teaching and research assistantships. These assistantships are sometimes part of financial packages provided to students or are pursued by the students independently. Some students seek other forms of outside employment. Many well-established forestry schools have endowed fellowships, scholarships and grants that are given to students based on merit and/or type of research project. A large percentage of these fellowships and grants provide funding for research, including equipment and supplies, but not for stipend or tuition reimbursement. Typically doctoral students receive more guaranteed funding upon acceptance into the program than do master's students.

Diversity of Scientists

The gender distribution of doctorates and doctoral students has moved toward representing the general population, but minority group participation in education leading to forestry or natural resource science careers has made little progress over the last three academic years.

Table 4-4 summarizes statistics about fall 1999 forest-science enrollment by gender, ethnicity, and citizenship (FAEIS, 2000). There were 764 doctoral students enrolled, 69.6 percent male and 30.4 female. Of the doctoral students, about 60 percent were U.S. Caucasians; 7 percent U.S. minority-group members; and 33 percent foreign nationals. Of the master's students about 82 percent U.S. Caucasians; 8 percent U.S. minority-group members; and 10 percent foreign nationals. Undergraduate students were overwhelmingly U.S. Caucasians (92 percent). The percentage of women was highest at the master's level (37 percent).

In 1996, 1997, and 1998, respectively, women in natural-resource programs earned 59, 67, and 82 (29, 29, and 33 percent), and in forest-science programs 34, 20, 30 (26, 17, and 21 percent) of the doctorates (FAEIS; 1997b, 1998b, and 1999b, respectively). These figures do not reflect the proportion of women in the population, or even in universities, but they do reflect a major change from only a few years ago, when very few women studied for doctorates in natural resources and forestry.

Table 4-4. Forest Sciences Enrollment Statistics by Gender, Ethnicity, and Citizenship, fall 1999.

Characteristics	Bachelor's		Master's		Doctoral	
	No.	%	No.	%	No.	%
Gender						
Male	5411	77.8	730	62.3	532	69.6
Female	1544	22.2	442	37.7	232	30.4
Race						
Caucasian	6420	92.3	964	82.3	458	59.9
Minority	503	7.2	92	7.8	54	7.1
African American	120	1.7	12	1.0	8	1.0
Asian	82	1.2	26	3.2	25	3.3
Hispanic	102	1.5	24	2.0	12	1.6
Native American	84	1.2	10	0.9	2	0.3
Unspecified	115	1.7	20	1.7	7	0.9
Foreign	32	0.5	116	9.9	252	33.0
Total	6955		1172		764	

Source: Food and Agriculture Education Information System (FAEIS 2000).

Enrollment figures suggest that the proportion of female doctoral graduates will increase. For the same 3 years noted above, enrollments of women in natural-resource fields were 483, 485, and 524 (34, 35, and 37 percent); in forest-science fields, the comparable enrollments were 191, 199, and 220 (26, 28, and 29 percent) (FAEIS; 1997a, 1998a, and 1999a, respectively).

Minority-group participation was considerably lower over the same three-year period with 12, 11, and 21 (6, 5, and 8 percent) natural-resources program graduates and 20, 14, and 10 (15, 12, and 7 percent) forest-science program graduates being members of identified minority groups. The number and proportion of minority group students enrolled in programs suggest improvement for natural-resource programs (114, 117, and 128; 8, 9, and 9 percent), but not for forest-science programs (73, 77, and 60; 10, 11, and 8 percent) (FAEIS; 1997b, 1998b, and 1999b, respectively).

In the statistics for both women and minority groups, it is clear that distribution across the various natural-resources and forest-science categories is highly skewed. Women graduates were most heavily represented in wildlife, environmental science and studies, and general forestry, and they were generally scarce in forest engineering, wood and paper products, forest soils, and mensuration and biometrics. Women students were most heavily represented in natural-resource conservation, environmental science and studies, wildlife, forest biology, and general forestry; and there were none in doctoral programs in harvesting and engineering or in hydrology. Minority group doctoral graduates were absent in most categories; the largest numbers were in forest management and environmental science and studies. Minority group students were represented best in natural-resource conservation, forest management, wildlife, and general forestry, but they were not represented at all in harvesting and engineering, forest hydrology, pulp and paper, and forest soils.

The statistics suggest an increasing proportion of women doctoral graduates entering the scientific workforce and a relatively static, and quite low, proportion of minority group graduates.

Future Demand for Scientists

Employment opportunities for scientists, engineers, and related specialists in agriculture, life science, and natural resources were summarized by Goecker, Gilmore, and Whatley (1999). The total U.S. employment of foresters and conservation scientists in 1996 was 37,000. The estimate of 43,000 needed in the year 2006 constitutes a 16 percent increase. Although many of these scientists might not seek work in research-specific fields, the increase in opportunity and demand for scientists reflects a substantial future demand for research scientists in forestry and natural resources. Currently, the demand for scientific professionals that support important federal missions has outstripped supply; it is imperative that scientists supplied by post-secondary education include all ethnic and gender groups at increasing rates if a strong science and technology workforce is to be ensured (National Science and Technology Council, 2000). Academic institutions will be challenged to educate these scientists to meet the demands, in terms of knowledge, number, and diversity.

INTERDISCIPLINARY AND INTEGRATIVE CAPABILITIES

The needs for integrative skills and interdisciplinary behavior in forestry and natural-resource science have been expressed in reviews that have been prepared over the last decade; the most prominent reviews are *Forestry Research: A Mandate for Change* (NRC, 1990) and *Sustaining the People's Lands* (Committee of Scientists, 1999). Meeting the challenges will probably require some components of individual doctoral curricula different from those traditionally considered. Multidisciplinary seminars, courses, and research and policy projects all might be useful. In addition, opportunities to interact substantively with doctoral students in other programs outside the normal seminar and classroom setting might be necessary (Box 4-2).

Opportunities exist for exposure to and thinking about integrative and multidisciplinary topics emerging in several universities, and the experiences that they offer might be particularly helpful. For example, multidisciplinary teams of graduate students and professors are working on projects to deal with issues. In other cases, seminars and courses are being taught by looking at important issues from a variety of perspectives in the humanities, and the social, managerial, and natural sciences. To the extent that these kinds of experiences are required of doctoral students, integrative and interdisciplinary awareness and ability might increase.

Institutional Arrangements

Several examples of successful federal programs represent innovative approaches to education and research and foster collaboration and diversification (Boxes 4-3, 4-4). These programs serve as examples of programs that could be implemented by USDA to improve disciplinary and multidisciplinary forestry education and research.

**Box 4-2 The Corporate Environmental Management Program (CEMP) at the University of Michigan:
An Example of Creative Partnerships
within the University and between Business and the University**

The Corporate Environmental Management Program (CEMP) is a joint-degree, three-year program between the Business School and the School of Natural Resources and Environment at the University of Michigan. CEMP students earn Master of Business Administration and Master of Science degrees. The program equips leaders, executives, and managers—regardless of whether they work in the private or public sector—with the skills and knowledge necessary to create environmentally and economically sustainable organizations.

In this program, students become well versed in both management methods and environmental sciences. In addition to classwork, the program includes executive education, summer internships, research projects, seminars by visiting practitioners, conferences on important environmental issues, and a lecture series on environmental management. Students in the program are supported in part by Weyerhaeuser Student Fellowships and General Motors Environmental Excellence Awards.

One example of such a success is the National Science Foundation (NSF) established Long Term Ecological Research (LTER) Network. The LTER Network was started in 1980 to support research on long-term ecological phenomena in the United States and has been extremely successful in its effort to facilitate collaboration among researchers. Over 1200 scientists and students investigating ecological processes over long temporal and broad spatial scales conduct research at LTER sites. Researchers are often associated with universities, but research teams also include members from the USDA Forest Service and other federal agencies. The LTER Network provides over 2000 ecological datasets available from LTER sites over the internet. The sites are models for how ecological research on forests can be conducted in a collaborative manner to improve understanding of ecological phenomenon (Box 4-4).

> **Box 4-3 National Science Foundation's (NSF) Integrative Graduate Education and Research Training (IGERT) Program**
>
> To meet the need for a cadre of broadly prepared Ph.D.s with multidisciplinary backgrounds and the technical, professional, and personal skills essential to addressing the varied career demands of the future, NSF created an agency-wide, multidisciplinary, graduate training program. The goal of the IGERT Program is to enable the development of innovative, research-based, graduate education and training activities that will produce a diverse group of new scientists and engineers well prepared for a broad spectrum of career opportunities. Supported projects must be based upon a multidisciplinary research theme and organized around a diverse group of investigators from U.S. Ph.D.-granting institutions with appropriate research and teaching interests and expertise.
>
> All IGERT projects are expected to incorporate the following features:
> - Vision, including goals and objectives, underlying an innovative program of graduate student training;
> - Comprehensive multidisciplinary research theme, appropriate for doctoral-level research, to serve as the foundation for training activities;
> - Training activities based on the integration of the multidisciplinary research theme with innovative educational opportunities;
> - Training environment that exposes students to state-of-the-art research instrumentation and/or methodologies;
> - Formal administrative plan and organizational structure that ensure the effective management of the requested resources to achieve the goals of the project;
> - Institutional strategy and operational plan for student recruitment , with special consideration to members of groups underrepresented in science and engineering, i.e., women, racial and ethnic minorities, and persons with disabilities, to ensure preparation of a diverse science and engineering workforce;
> - Well-defined strategy for assessment of project performance.
>
> In the two-stage IGERT competition, applicants first submit a preliminary proposal (preproposal) that outlines the planned IGERT activity; in the second stage, invited applicants submit a formal proposal. Invitations to submit a formal proposal are extended on the basis of merit review of the preproposals; only invited formal proposals are accepted.

> **Box 4-4 NSF's Luquillo Long-Term Ecological Research (LTER)—An Example of Forestry Research Conducted through a Creative Partnership between Universities and Federal Research Agencies**
>
> The Luquillo Experimental Forest in the subtropical wet forests of Puerto Rico was established in 1989. There are presently about 40 scientific researchers working at this LTER site. Researchers working in the Luquillo LTER are affiliated with about 20 universities across the U.S., non-profit and private research organizations, the U.S. Fish and Wildlife Service, U.S. Geological Survey, and the USDA Forest Service's Forest Product Laboratory, Caribbean National Forest, and the International Institute of Tropical Forestry. Researchers working at this site have produced over 450 peer-reviewed articles over the last 10 years with the large majority resulting from collaborative projects.
>
> Shortly after this LTER was established, Hurricane Hugo passed over the island. Since that time, the researchers there have been studying the effects of disturbance on the structure and functioning of the system. The variety of expertise among the researchers who work at the site has permitted successful studies of how disturbance affects components of the communities and ecosystems in the study forest. For example, the Hurricane Recovery Plot, a 16-ha study area at El Verde Research Area, was established shortly after the hurricane passed, to monitor changes in vegetation composition. Other researchers have monitored changes in amphibian, lizard, shrimp, and snail populations as well as changes in plant productivity, leaf litter decomposition rates, and other ecosystem processes. The research resulting from the Luquillo LTER site has been effective in changing thinking about the role of disturbance in systems and has helped ecologists understand how integral disturbance can be community and ecosystem dynamics.

CONCLUSIONS AND RECOMMENDATIONS

Our discussion of the many facets of professional forestry education leads to several conclusions and recommendations about what might be done, and what might be done better, to enhance of our forestry-research capacity.

Recommendation 4-1

University programs should assume a renewed commitment to the fundamental areas of scholarship and research related to forest sciences that have diminished in recent years, and should adopt an enhanced, broad, integrative, and interdisciplinary programmatic approach to curricula at the graduate level.

Basic fields—including field biology, population genetics, plant systematics, and plant taxonomy—are fundamental to understanding any biologic system. All too often, faculty, support staff, and their facilities in such fundamental fields as genetics, physiology, pathology, and entomology have been allowed to decline in universities and natural resource agencies. The intellectual capital in many of these fundamental fields is dangerously low, and this lack of capacity will affect the nation's ability to implement new programs of research and development.

We need to consider developing curricula that include more mixing of students from various disciplines through seminars, capstone courses and experiences, and the use of multidisciplinary teams in teaching. In the future, teams of scientists from multiple disciplines will carry out much of forestry research, and this requires team behavior. The primary implementation problem is to capture enough time in already crowded curricula and teaching schedules for "mixing" activities. If school-wide or department-wide cores can be designated to include these activities for all students, with specialties viewed as additions to the common core, the "room" problem is solved by reducing the time allocated to specialization.

At the same time, all students need to be introduced to the methods and processes of science. Multidisciplinary teams work best if all members have a strong foundation in science. Thus, "research methods" classes cutting across disciplines should introduce students in various disciplines to specific approaches to science and should enhance disciplinary "cross pollination" among students. Implementation here requires "only" the addition of a course designed for all specialties. The usual research methods course focuses on the preparation of written study plans. This can double as doctoral dissertation or master's thesis prospectuses, or they can be research- grant applications.

The body of skills developed in scientific education would not be complete without enhancing communication skills as a core professional attribute for doctoral students. The "mixing" activities mentioned above help students to improve their communication skills by requiring them to explain, in a reduced-jargon environment, what they are doing and why they are doing it. In addition, specific communication courses might be offered for graduate and research students. Often, these can be integrated with, or parallel to, research methods courses. All courses should stress communication that allows spanning disciplines in writing and speech.

Students need to be exposed to a formal "systems" approach that can be useful in organizing graduate curricula and research. In addition to the offering of formal systems courses, such as ecology, "systems thinking" should be embodied in teaching and learning through the use of examples in which the description and integration of systems components are demonstrated. The systems approach can be enhanced if we ensure that all future researchers have a core of science method, a specialization in which they have competent depth, and an appreciation of a wide array of other disciplines, including enough of their specialized languages to communicate effectively with people working in them. Thus, breadth and depth should be considered compatible in graduate programs. That principle suggests three universal curricular components: (1) core knowledge of science and its processes; (2) a specialization that confers complete currency in a field; and (3) experience through courses and other interactions that confers an awareness level of competence in several fields of natural-resources research.

Success in meeting those needs can be reduced by increased faculty specialization and intensified competition in subfields for money and recognition. Managers of academic programs must be aware of the time and resources necessary to support synthesis and cooperative efforts by faculty and students. Students must be presented with observable evidence that a core knowledge of science and work among disciplines "pays off" and that these must be added to, rather than replace, a specialty. Students will do this only in so far as their role models on the faculty are seen to pursue this course successfully and that will likely require retraining many faculty members.

Recommendation 4-2

Universities should develop joint programming in geographic regions to ensure a "critical mass" of faculty and mentoring expertise in fields where expertise might be dispersed among the universities.

There are a wide variety of subfields in forestry and natural resources, and few institutions can produce doctoral graduates in many subfields. Regional cooperation might be viewed as a way to expand capacity by pooling resources in important areas. The building of regional coalitions among universities for the purpose of graduate education could enhance the education of students and lead to cost-effective expansion of the capacity to develop forest and natural-resource scientists.

Universities, government, industry, and private groups should work toward innovative and creative partnerships to a much greater extent than in the past to ensure that the spectrum of forestry education, research, and development interests is covered (Box 4-2). Each organization should play a unique role. The unique opportunities offered by each research entity should be better identified, and mechanisms for coordinating across institutional niches should be better developed. Furthermore, each institution, or consortium of institutions, should concentrate its research capital in specific (and perhaps limited) fields of forestry research where it operates best or has some recognized institutional advantage. One of the ways to increase cooperation is to bring federal, state, and private sector scientists into the academic fabric where needed to augment the expertise of university faculty in preparing future scientists. Collaboration of non-university scientists in the academic fabric could expand the critical mass of scientists and educators preparing future scientists.

5

Capacity of Forestry-Research Organizations to Meet Future Research Needs

Four factors are important in enabling a research organization to perform useful research and meet future research needs. They are:

- Continuity through time allowing for adequate and consistent resources to maintain and improve operations,
- Availability of up-to-date facilities and equipment,
- Access to skilled and competent scientists, managers, and staff, and
- Focus on high-priority goals and needs.

Our nation's forestry-research engine has made substantial progress over the last several decades, but it appears to be struggling at some level with respect to all four factors listed above. It is difficult to address those factors separately, because they are integral to each other. For example, without adequate and consistent human and financial resources over time, it is impossible to maintain quality researchers, programs, facilities, and equipment. The declines and trends described in Chapters 3 and 4 in scientific, educational, and fiscal resources that make up the U.S. forestry-research enterprise are cause for concern. This chapter summarizes the concern and presents information in the context of the capacity of forestry-research organizations to meet future research needs.

CONTINUITY THROUGH TIME: RESOURCES TO MAINTAIN OPERATIONS

The forest industry's contribution to the domestic gross national product and the number of people employed directly and indirectly by forestry are large, and forestry research efforts and support should be large to maintain them. When the National Forest system, Bureau of Land Management, and the National Park Service are considered, an even stronger case is made for the importance of forestry research to our nation. Given the high cost of modern research in biotechnology, genomics, and ecosystems, the need for adequate support of forestry research is even greater.

Chapters 2, 3, and 4 of this report provide recommendations for addressing deficiencies in scientific and fiscal resources needed to secure our nation's future forestry-research capacity. The recommendations encompass university, government, and industry. Implementing some of the recommendations might require new federal funding, which is often difficult to obtain. The search for new funding will continue, but lasting change might occur best through reshaping and development of important new models and systems to generate the dollars for research. Federal, state, and local law and regulatory changes could be made to encourage investment in forest research.

Models such as cooperative university, industry, and federal research appear to be functioning well and should be considered (Box 5-1). For any such models to function three things conditions are necessary: stakeholders must agree that there is a need, there must be an equitable system to secure the required funds; and there must be a defined process to set priorities and allocate the funds.

Box 5-1
Northwest Stand Management Cooperative (SMC)

The mission of the Northwest Stand Management Cooperative (SMC) is to provide a continuing source of high quality information on the long-term effects of silvicultural treatments on stand and tree growth and development and on wood and product quality. The SMC is composed of 19 forest industry members; six state, provincial, and federal agencies; three suppliers; and four universities. The Policy Committee, composed of dues-paying members, controls policy and establishes goals with the aid of the Technical Advisory Committee in silviculture, nutrition, wood quality, and modeling.

The SMC annual budget over the last five years has ranged from $0.9 to $1.1 million; 60 percent comes from member dues, 20 percent from grants and contracts, and 20 percent from institutional members in the form of salaries, facilities, and administrative support. The SMC database represents 435 installations containing 4427 plots, with data on a quarter-million trees in Washington, Oregon, and British Columbia. SMC is headquartered at the College of Forest Resources, University of Washington, which provides administration and staffing.

University System

The university system is struggling to find the resources to maintain high-quality forest research and in general is falling further behind. Facilities and equipment for the most part are out-of-date (National Science Foundation, 1996). This makes it difficult to draw good students into forest research and to provide the training required for science. Many faculty also find themselves operating mostly on soft money, which can make it difficult to maintain a focus on specific priorities.

Current administrative structures and funding mechanisms do not provide the "critical mass", appropriate organization, and focus to meet many of the nation's needs in forestry research. To implement the research priorities outlined in this report, centers of excellence in forestry are proposed. These would provide a new mechanism for focusing substantial effort directly on specific research needs for practical applications and basic science in forestry (Box 5-2). The value and efficiencies of focusing many scattered research facilities by establishing "centers of emphasis" has been reported previously (University of Idaho, 1983; National Research Council 1990; National Research Council, 1995).

As previously recommended, centers of excellence might be a means of accomplishing interorganizational research cooperation. The centers could help to institutionalize cooperation among organizations using existing scientists or hiring new staff that would administer joint programs. The centers usually are housed at particular organizations, but seek cooperation among many partners and funding from outside sources. Centers could add another layer of supervision for scientists, so careful thought needs to be given to administrative responsibilities associated with the centers.

Box 5-2
Centers of Excellence in Forestry

The complexity of contemporary forest-resources research issues requires a diversity of expertise that is seldom found in any single institution or organization. The concept of a center of excellence in forestry is that researchers from different organizations could interact effectively, by using modern communication technology, to address complex high-priority research issues without being in the same location.

A center could involve any combination of university, industry, and government participation. Centers could be coordinated by existing personnel or by the funding of new administrative positions; in either case, the organizations involved would need to provide administration and leadership in securing funding, coordinating research activities, and disseminating new information.

This concept calls for something similar to research foundations at most universities and to research centers established by foundations, but the centers of excellence would be more focused. A center would be a variation of the existing National Science Foundation science and technology centers. The major advantages of the new concept are the creation of an entity possessing the diversity of expertise required to solve complex contemporary research problems, enhanced cooperation among scientists in different organizations, use of the best scientific expertise, ability to attract sufficient resources to address large-scale research issues, and flexibility.

"Virtual centers" might be a means to achieve new cooperation among various partners without much added administration. The centers could continue to rely on existing scientists and programs but try to seek synergies and fill gaps at participating institutions to achieve well-defined research goals. These cross-institutional centers would need new funding to provide an incentive for collaboration but might offer the promise of more focused cooperation without greater administrative overhead. If virtual centers remain modest in size or scope, they might be administered by management teams in existing organizations. If large external funding is received and numerous projects initiated, a formal center director and administrative structure will probably be needed.

Munson (1999) concurred that research problems in the future will usually need to be addressed in teams rather than in the classical single-investigator model. Research initiatives will often be organized at the regional level with cross-organizational structures. No single institution could or should dominate the research agenda. Intellectual property rights will be more highly valued by research sponsors and researchers, and business arrangements to protect those resources will be developed. Last, fixed-term research agreements focused on cooperative arrangements will be increasingly important, and long-term permanent projects will decrease in importance.

It is expected that centers of excellence would have a large education and training component associated with the research focus. Such centers could be virtual centers with a limited funding period. Projects would be periodically reviewed and targeted to specific research objectives. Funding would be competitive, and projects would be focused on critical needs in forestry and forest sciences. Substantial involvement would be expected from industry, government, universities, and forestry-related NGOs for review of proposals and management of projects. Two examples of virtual centers at work are the cooperative ecosystem studies units and the Valuation of Wildland Resource Benefits project (Box 5-3).

Competitive funding remains an excellent approach to ensure that research expenditures are used for high-priority needs (National Research Council, 2000). Funding for new competitive grants is difficult to obtain, but the case must be made on the basis of the importance of the collective needs of resource production and protection.

Competitive funding has many desirable attributes but has a downside that any scientist working on soft money knows well. Accessing the competitive grant pools requires considerable time for proposal preparation and submission, which often fail to result in funding. Generally, less than 10 to 25 percent of proposals submitted to NSF, NRI, and Agenda 2020 have been funded. This has a great effect on the quantity of research that can be accomplished by organizations that depend heavily on competitive grant money. The present process increases the quality of science performed, but the value of spending so many scientist-years in securing the resources to carry out research should be weighed against other models that accomplish a similar outcome. For example, reducing by one-third the time spent on proposal development and winning grant applications could effectively free up the equivalent of many scientists per year for research without adding costs.

> **Box 5-3**
> **Virtual Center Concept at Work**
>
> *Cooperative Ecosystem Studies Units*
>
> The federal and university partnerships formed through cooperative ecosystem studies units (CESUs) are a relatively new way to provide research, education, and technical assistance for the benefit of both agencies and universities. CESUs bring together managers, scientists, and educators in a national network overseen by a national coordinating committee and regional CESU committees. Ten regional CESUs cover the Chesapeake Watershed, Colorado Plateau, Desert Southwest, Great Basin, Great Plains, North Atlantic Coast, Pacific Northwest, Rocky Mountains, South Florida and the Carribean, and the Souther Appalachian Mountains.
>
> regions. These units are designed to identify programs that should be pursued, to enable the sharing of resources and funds without a lot of red tape, and to facilitate crossing institutional boundaries. The Rocky Mountain CESU is managed by the University of Montana with participation of the University of Idaho, Montana State University, Salish Kootenai College, Utah State University, and Washington State University and with, as federal partners, the Bureau of Land Management, National Park Service, USDA Forest Service, and the U.S. Geological Survey-Biological Resources Division. An Executive Committee of all partners oversees the management of the unit, and a Manager's Committee provides advice on programmatic themes and directions.
>
> *Valuation of Wildland Resource Benefits*
>
> An example of successful management and brokering of collaborative research activities is the long-standing policy of the Forest Service's Valuation of Wildland Resource Benefits project (Rocky Mountain Research Station) to facilitate the work of many university scientists studying recreation and other amenity resource valuation issues. This project routinely facilitates the work of university scientists coast to coast, overseeing many studies that fit together as pieces for understanding valuation of wildland resources.

Another method of increasing resources focused on forest research would be to create innovative risk and reward processes that encourage reallocation of existing resources in allied areas and organizations (for example, wildlife, hydrology, genomics, and geographic information systems or GIS). If a new model can be developed and implemented that allows research scientists to come together and focus on high-priority subjects while minimizing proposal preparation, it should be possible to increase scientist productivity. The challenge in defining such a system will be to keep it simple.

As discussed in Chapters 3 and 4, forestry research needs to be better integrated with development and extension to ensure dissemination of research results. University cooperative extension programs offer the promise of achieving that objective, but for various reasons they have not realized their potential. Much of the failure can be attributed to modest funding. With only about $20 million for all federal and state contributions, forestry efforts pale in comparison with agricultural. In addition to direct funding, financial incentives are needed to draw researchers and technology-transfer experts together in designing research programs and distributing their results (National Research Council, 1998).

Interdisciplinary and interinstitutional efforts mean that scientists not only must be trained in a technical skill, but also must be trained in skills that allow them to work in complex teams focused on common goals. Many scientists find that difficult. Reward systems at individual universities—such as tenure, promotion, and pay—must encourage cooperation and extend across institutions. Present reward systems tend to work against a cooperative model, instead favoring and rewarding individuals. A system that encourages both without stifling individual creativity is desired.

Forest Industry

The private sector has been unable to maintain continuity of funding. When inflation is taken into consideration, even companies with large forest research organizations have become smaller with periodic budget cuts and redesigns. Industry funding for sustainable forestry research totaled $68 million in 1999, up from $60 million in 1996 (see Table 3-9), but most of this spending occurred in only four companies, and one of the challenges for the industry is to engage the majority of the industry.

Ellefson and Ek (1996) estimated that private, forest products research in the United States amounted to almost $900 million. Private forestry R&D expenditures were about $60 to 70 million. The balance of $830 to 840 million of private research was focused mostly on proprietary forest products and paper science R&D. In 1991, forest products R&D expenditures amounted to less than 1 percent of industry's domestic sales—0.8 percent for paper and allied products and 0.7 percent for lumber, wood products, and furniture. One might expect that forestry R&D expenditures are even less, given their small proportion of total forest products R&D expenditures. The proportion of forest products expenditures for R&D was much less than the average for all industries, which was 4.7 percent of domestic sales, and only one-tenth as much as that of computer science.

Industrial research has tended to focus on projects that have near-term effects rather than longer-term projects that pay off in the future. Ellefson and Ek (1996) found that only 8 percent of the wood-based industry's R&D were aimed at developing fundamentally new knowledge. There are notable exceptions in biotechnology, in which major investments were made by several of the major forest products companies.

The forest industry has led in the creation of the Agenda 2020 program. That effort is developing a new model to focus new and old money on high-priority research needs of the forest products industry. The model requires the industry to provide a minimum of 25 percent of the funds for the research and is aimed at pre-competitive research; thus it tends to focus on longer-term projects. It encourages cross-agency and cross-organizational participation and has developed a total portfolio of $13 million for sustainable forestry projects in its three years of operation. Both the Department of Energy (DOE) and the Forest Service are active partners in the program. In addition to forestry, the program has many grants related to forest products and pulp and paper processing.

The industry also encourages and participates in consortia and research cooperatives with universities, state and federal agencies, and nongovernmental organizations. For example, the Forest Biology Research Cooperative and the Biotechnology Consortium were formed, in 1996 and 1997 respectively, with the University of Florida School of Forest Resources and Conservation, industry, and the U.S. Forest Service. A cooperative is usually centered at one university and has industrial, other university, and often state and federal forestry organizations as members. The cooperatives usually have narrowly focused purposes, such as tree improvement, growth and yield, or vegetation management, and they follow study plans determined annually by their members. North Carolina State University maintains several cooperative research programs with forest-based industries that address research designed to maximize the productivity of commercial timberland, maximize economic efficiency and protect the environment. Most co-ops have had fixed dues for all members, regardless of company size. Overall, forest management cooperatives have expanded slowly over the last five decades, and have had excellent success at targeting specific research needs. Forest and related industries now contribute about $10 million to cooperatives in the United States.

Consortia tend to have broader purposes and more loosely organized structures. The western Stand Management Cooperative—which examined growth and yield, silviculture, and wood quality—typifies such consortia. This cooperative effort has members and studies at universities and organizations throughout the West. A similar approach, the Southern Forest Resource Assessment Consortium (SOFAC), was formed to study timber supply issues in 1994. SOFAC has had about 15 forest industry, consulting, and state member organizations that pay annual dues and has had several USDA Forest Service research work unit contributors. The consortium is administered through the Southern Research Station and has funded timber supply research and modeling projects at seven southern universities.

Clearly, the trend in industry is to do more targeted outsourcing of its research needs via cooperatives, universities, or other research organizations and to spend less internally. In this age of consolidation among forest products companies, many of these cooperatives might find it difficult to fund their programs with the same fixed dues for all members as the number of member companies declines. This suggests that the method of assessing membership fees might need revision.

USDA Forest Service

The USDA Forest Service has experienced fluctuating budget levels and budget erosion because of inflation. Despite increases in Forest Service forest management research funding, the forest industry perception is that attrition and retirement have dramatically reduced the skills and competencies that are critical to intensive forest management while skill areas aimed at ecosystems have increased. Present budgets cover salaries, fixed operating costs, and overhead but leave little for new equipment, the variable costs of carrying out research, and cooperative research. Recall that Table 3-3 presents trends in research skills and staffing in the Forest Service.

Continuity through time, facilities, scientific and managerial talent, and strategic directions are the key determinants of Forest Service research success, as in all organizations. Despite substantial reductions in scientific staff over the last two decades, the Forest Service research team remains exceptionally strong. That is evidenced by its thousands of publications each year; leadership in major regional and national studies; active involvement in local, national, and international professional societies; and a host of other criteria. No other forest research organization in the world can focus more resources on important problems or have such a large resource management organization that can be directed to examine major social issues.

FACILITIES AND EQUIPMENT TO PERFORM HIGH-QUALITY RESEARCH

The ability to carry out high quality, productive research requires an up-to-date physical plant and equipment. *Forestry Research: A Mandate for Change* (NRC, 1990) stated that the physical plant and equipment at many forestry research stations and forestry colleges were inadequate. Since its publication in 1990, funding has been even less adequate to keep pace with changing technology. With the exception of a few forest products companies, industrial research laboratories also have not kept pace with technology needs. Industry consolidation continues to play a role in reducing the number of research laboratories and researchers focused on sustainable forestry targets. The same can be said for the university system: most institutions are underequipped, understaffed, and underfunded.

Without an adequate physical plant and up-to-date equipment, it will not be possible to do the research required to address society's forest research needs. It will be difficult or impossible for our university system to train and educate new scientists with the necessary skills. The disparity between highly capitalized and efficient private sector research and often "shoestring" public facilities is widening. Investments in forestry-research equipment and physical plants are less than in agriculture, information technology, pharmaceuticals, or medicine. These shortcomings of facilities and equipment handicap our ability to perform research and solve pressing forestry and natural resource problems (National Science and Technology Council, 1999).

The issue of facilities and equipment involves how research is performed and who should be responsible. A recent report by the Strategic Planning Task Force on Research Facilities (USDA, 1999) examined "current and planned agriculture research facilities, funded in whole or part by federal monies, to ensure that a comprehensive research capacity is maintained." The task force that prepared the report came up with many findings and included Forest Service research facilities. Its recommendations have a bearing on our study. In its executive summary, the task force stated (P. v):

Underpinning the Task Force's vision are four basic propositions. First, the bedrock philosophy for creating research capacity is quality scientists who are well educated and well trained. Second, investments in research infrastructure are driven by a philosophy of maximum flexibility and collaborative use of laboratories and equipment. Third, the laboratories in the broadly defined food, agriculture, and forestry-research systems are appropriately connected and, to the extent possible, create linkages resulting in improved productivity—a major change from earlier emphasis on physical structures. Fourth, the public at large has timely and equal access to research results produced by this system.

The task force determined that research facilities should be classified into three types of responsibility:

- Uniquely federal—responsibilities singularly proper for the federal sector,
- Appropriately federal—responsibilities suitable for the federal sector and shared with other sectors (universities, other research organizations, and private sector),
- Not uniquely or appropriately federal—responsibilities not fitting the federal sector.

As the research propositions and classifications suggest, the Strategic Planning Task Force on Research Facilities espoused a careful examination of research facilities and infrastructure with the intention of maximizing returns to federal investments via integrated, interconnected research facilities. The task force (USDA 1999) concluded that much of federal research objectives could actually be accomplished with fewer facilities but through cooperation with external partners (P. 9):

> The Task Force urges the intramural agencies to concentrate their efforts on facilitation, and, if appropriate, funding of major mission-oriented research and development programs with specific output expectations wherever those programs can be best accomplished within the vast extramural research capacity. The Department of Agriculture should not focus on carrying work only in federally owned facilities. Instead, the focus should be on funding the work and ensuring that results come from the best sources available at universities, private institutions, and industry—thus establishing virtual facilities. This approach should not be interpreted to mean that intramural research agencies will lose control of the funding, face reduced budgets, or be restricted in the influence they exert over research. To the contrary, this approach presents an opportunity for the intramural research agencies to manage their resources in whichever ways support the largest number of first-rate scientists and produce the most meaningful outcomes and outputs.

Resources could be concentrated in such collaborative or virtual research facilities, which could be supported by new modern communication technology. Despite their limitations, the Forest Service facilities and equipment are still much better than those of many universities. Only a few forest-products firms and universities have made substantial forestry-research facility investments that are not matched by the Forest Service. Funds and scientists—from federal agencies and from other public and private partners—could be used more effectively where the federal role is not unique. That would allow a greater "critical mass" of scientists from multiple disciplines, collaboration among laboratories in the federal and non-federal research sectors, and co-location of federal laboratories with other laboratories or universities, where practical, to be most effective. The task force vision applied explicitly to Forest Service research facilities and has broader implications than for facilities.

ACCESS TO PEOPLE WITH APPROPRIATE SKILLS AND COMPETENCIES

As the report by the Strategic Planning Task Force on Research Facilities (USDA 1999) stated, high-quality scientists who are well educated and well trained are the bedrock of creative research capacity. Chapter 4 of the present report summarized the status of the education of our future researchers and called for improvements in education. This section of this chapter focuses on the ability of scientists to perform the required forestry research in the public or private sector.

Research projects are increasingly complex and interrelated and that makes cross-functional and interdisciplinary teams necessary. Mechanisms must be put into place to encourage cross-organizational, cross-functional, and cross-geographic interdisciplinary teams to address the increasingly complex issues. Examples of the complexity are seen when we recognize the interactions among physiology, biochemistry, and biotechnology and the multiplicity of interactions between and within ecosystems. Biological complexity requires high quality research to understand *why* trees and plants respond, as opposed to simply empirically measuring responses. Our capacity to do this type of fundamental research is questionable. However, we will find it harder and harder to understand the nature of the systems we manage unless we address the issue.

We will also find it more and more expensive to do research that addresses those questions only in an empirical way. Teams of scientists with different skills must carry out such research, and it is improbable that a single organization will contain all the skills necessary. Examples abound in the computer industry and the biotechnology industry. Addressing complex outcomes requires cobwebs of interactions, cross licensing, and alliances, joint ventures, and collaborations. In addition to the biological complexity, scientists are increasingly asked to address the socioeconomic impacts of implementing research. That further increases the complexity of their interactions and requires interactions with social and economic researchers.

Federal and university research scientist capacity has declined in the last decade. Despite their large size, Forest Service research support and research capacity have decreased steadily for years. Declines in numbers of scientists and budget allocations have decreased the foundations of research. Forestry faculty numbers have remained fairly stable, in comparison with the Forest Service personnel reductions. Given their existing personnel and infrastructure bases, more financial support to augment existing resources could greatly increase the Forest Service's (and other research organizations') research capacity and ability to perform and deliver forestry research directed at high-priority needs. Such support, however, must be contingent on better use, cooperation, and collaboration among scientists at least in the public sector, and as appropriate in the private sector.

FOCUS ON HIGH-PRIORITY GOALS AND NEEDS

We will accomplish little if we establish a well-funded research system that is not well guided and focused on national priorities. In industry, focusing on business goals aligns cross-functional organizations. Such a system is described in detail in *Third Generation R&D* (Little, 1991). The process by which national goals are agreed on and used to align R&D resources is critical.

Identifying and focusing on high-priority research needs is the largest issue regarding the ability of the Forest Service to perform and deliver fundamental and emerging forestry research. The Forest Service has had difficulty in identifying and articulating a strategic vision for forestry research, is challenged to execute a strategy, has earned a reputation among some of its clients for weak communication, and has often become embroiled in political controversies.

Such political controversy has in some cases eroded the ability of the Forest Service to perform high-quality research. Such projects as the spotted owl controversy and the president's plan for the Pacific Northwest are contentious, and no solutions are likely to placate diverse interest groups with inherently different values. The 1999 Committee of Scientists proposal suggesting that Forest Service researchers review and comment on the science base of national forest plans could further dilute the credibility and independence of the agency and as detract from scientists' time for scientific research.

It might be easy to attribute the Forest Service's difficulties in defining vision and direction to external forces, such as its difficult political operating environment. However, the agency itself needs to bear more responsibility for its problems and their fate. Whether because of external forces or internal resistance, Forest Service research at the national level does not appear to have a vision for its future. With the exception of strategic planning at the Station level, the last and very modest national strategic planning exercise for research by the Forest Service occurred in the 1980s (USDA Forest Service 1990). The recommendations and priorities defined in Chapters 3 and 4 of the present report provide a starting point for developing a comprehensive vision to ensure research capacity.

Furthermore, the agency might benefit from increasing communication efforts with partners and clientele. Some interest groups that interact with the Forest Service perceive that their needs are not being met. For example, the environmental community perceives that its concerns are not being addressed adequately. Fish and wildlife and social science researchers perceive that although the Forest Service has some excellent researchers, the agency provides only a token effort in those disciplines. The forest industry often perceives that its requests for research aimed at intensively managed forests have been ignored. In response to such criticisms, the agency holds meetings and responds with discussion but seldom changes allocation of resources or research work unit missions.

The Forest Service's deficiencies in strategic directions could be contrasted with what one expects to be the more disparate efforts of the nation's professional forestry schools. Despite their diversity and inherent independence, the schools have led in developing not only an internal strategic planning document and teaching, research, and extension plan (NAPFSC 1998), but also an effort to do the same for nonfederal lands (NAPFSC and CSREES 1999). A vision is needed for the Forest Service, and it should be developed in cooperation with its traditional and new partners. Industry, through its partnership with DOE in Agenda 2020, has produced a focused research agenda. Other agencies, such as EPA and NASA, seem more interested in particular components of forestry questions that are related to their missions and might pursue them purposefully or opportunistically.

CONCLUSIONS AND RECOMMENDATIONS

Drawing from the continuum suggested by Roussopolous (1999) and described in Chapter 1, what research organizations are appropriate today? The spectrum of forest industry, Forest Service, environmental agency, national science, academic, and nongovernment research organizations seems to cover the gamut of possibilities. However, not all of the variants have been applied in forestry, and creativity in new approaches with existing organizations has merit.

Most agree that we must do more research with fewer resources, we should collaborate more on projects of mutual interests, and we should take a broader perspective in our research. In general, that consensus provides considerable basis for recommendations. However, we note that our current research structures were based on decades of incremental improvement, and we do not recommend casting them aside as much as modifying them. Burkhart (1999) pointed out that some research duplication is not only useful, but also necessary to accelerate and validate progress.

Current research organizations have merits, but we need to move toward new systems appropriate for new social and political environments. Existing resource management organizations must cooperate better, and partnerships that improve on unilateral research possible by single organizations must be formed. Research cooperatives and research consortia are one evolving means of developing research synergies. Research consortia provide a means for broader cooperation among more partners—universities, industry, and states, federal, and nongovernment organizations. However, creation of centers focused on specific research emphasis that involve many players is a need that continues to grow as forestry research continues to broaden and demands continue to expand.

Recommendation 5-1

Centers of excellence in forestry should be established and administered by USDA. These programs and awarded projects should (1) support interdisciplinary and interorganizational activities, (2) focus on increasing minority student participation in education and research, (3) clearly justify how new forestry-research approaches and capacity will be enhanced, and (4) undergo initial and periodic review.

Establishing centers of excellence in forestry for fields related to forestry research and education will require investment. The magnitude of investment will depend on the type of centers established. As noted in by the National Research Council in 1990, the centers need not be "bricks and mortar." Options for "virtual" centers described in the current report address the need to work within the existing structure and fiscal constraints. Regardless of the type of center established, focusing research efforts and increasing efficiency of existing resources through centers will result in enhanced research and education. The goals of centers of excellence in would include: (1) working closely with government agencies and other organizations to develop new research and education collaborations and partnerships; (2) encouraging and providing opportunities for university faculty and government researchers to conduct integrated interinstitutional research; (3) providing incentives for minority group students to enter and remain in forestry research; (4) establishing measurable program goals and objectives; and (5) developing and implementing evaluations to assess the effectiveness and outcomes of programs and financial performance.

Effective recruitment and outreach run by universities and governments are essential for reaching all sectors of society. However, such programs in forestry education and research have been largely ineffective in increasing minority representation over the last several decades. Minority group participation in science education, graduate-level training, and forestry teaching, research, and development is inadequate. Recruitment and outreach need greater attention and resources. Support is broadly needed to enhance minority group participation in forestry research, but a portion of it should be targeted at topics identified in Chapter 2 as needing particular attention. Achieving an ethnically and racially diverse group of forestry scientists will require extraordinary recruiting efforts. Support of such students through awards provided through centers of excellence in forestry is one key factor in ensuring a better prepared and more diverse research workforce in the future.

Funding of university research is a concern, with limited resources. Competition among universities and public entities encourages better research and faster dissemination of results. It also provides replication of results, greater confidence in results and wider applicability. Furthermore, the competitive grant process is almost universally revered for inducing the best scientific proposals and research. The wide geographic distribution of recipients of competitive grant funds has enhanced political support for increases in federal funding. More competition in forestry research has merit, including in research agencies that now have their own forestry-research portfolios. As described in Chapter 3, we need a mix of competition to generate new ideas and accelerate progress, solid base funding to support personnel and infrastructure, and collaboration to ensure that scarce research funds are used wisely.

Direct grants programs are another means of advancing forestry research. Grants provide a means to set specific scientific objectives and then seek proposals and projects to accomplish the objectives. That allows a large amount of scientific creativity, although it tends to be somewhat weak in monitoring and modifications or in accomplishing planned results. Traditional requests for proposals (RFPs) have focused on single institutions or even single-investigator research. There already has been a substantial movement to broaden this base, and it needs to continue.

Broad, long-range programs, such as NSF's Long Term Ecological Research (LTER) Network, provide another means to achieve integrated research projects at a single location but with cooperation among many organizations. They ensure that many partners participate in the research, and they focus research on questions of broad interest more than do individual projects developed across the landscape.

Recommendation 5-2

Clear federal research facility mandates—such as long-term ecological research sites, experimental forest and natural resource areas, and watershed monitoring facilities—should receive priority for retention and enhancement, and a system of periodic review of all facilities should be implemented and maintained.

The LTER Network exemplifies one mechanism for enabling valuable research and creating needed capacity, ideas endorsed throughout this report. The LTER network has been successful in: collecting scientific data on ecological phenomena over long temporal and large spatial scales, creating a legacy for such research, facilitating collaborating among researchers from diverse geographic locations, conducting major synthetic projects, and in providing easily accessible data for researchers. These research sites would benefit from periodic external review to ensure that they achieve their original objectives and investigate appropriate new subjects.

6

Summary and Conclusions

Forestry education and research may be classified in various ways, and new disciplines evolve slowly but distinctly over time. The committee used input from a public workshop on forestry research capacity and a review of available forestry strategic planning documents to develop a broad list of important foundation and emerging research needs and disciplines. Those important foundation forestry research needs are biology, ecology, and silviculture; forest genetics; forest management, economics, and policy; and wood and materials science. The important emerging research disciplines are human and natural resource interactions; ecosystem function, health, and management; forest systems at various scales of space and time; forest monitoring, analysis and adaptive management; and forest biotechnology. To some extent, these broad priority areas for forestry research and education reflect those identified in *Forestry Research: A Mandate for Change* (National Research Council, 1990). But their implications and applications go much further given the dramatic changes we have seen in Sustainable Forest Management, certification, biotechnology, social change, and other factors that affected forestry in the 1990s.

The identified broad priority areas for forest science education and research suggest that we need continued focus on foundation or traditional areas of forestry research, as well as new foci on emerging areas of research. The term foundation suggests that we cannot just jump to emerging areas, or neglect traditional areas, because new disciplines have enduring needs that are based on evolving knowledge about biology, ecology, forest management, measurements, policy, and other forestry research and education disciplines. For example, complex forest health monitoring issues still revolve around basic information about tree physiology, response to nutrients or pathogens in the atmosphere and soil, pest/host interactions, and climate. These foundation areas require continued focus, people, and funds for scientists to improve their understanding of basic biophysical processes, and to make sound recommendations for forest management and protection.

The emerging research and education disciplines identified in this report are evolutionary, not revolutionary, extensions of the foundation research areas. On one hand, forest biotechnology is the extension of the focus on trees or plants at the cellular level to the level of DNA and genetic properties and markers. These efforts promise to revolutionize production of trees with desirable characteristics for commercial purposes, and precise identification and preservation of biodiversity at the most basic level. On the other hand, tree, plant, wildlife, water, and soil taxonomy and interactions at the stand level are being extended to examine ecosystems, human interactions, landscape effects, and adaptive management. These broad views of integrative natural resource and human resource management and impacts are closely reflected in the recent development of Criteria and Indicators for Sustainable Forest Management and in industry and environmental forest certification approaches.

RECOMMENDATIONS

The scientific capacity in forestry research and education in the United States is at risk. This report identifies many encouraging facts regarding the extent and diversity of forestry research capacity. But the status quo of incremental changes in vision, funding, cooperation, and staffing will lead to diminished, not enhanced, research, education, and practice. The effects of reduced research and education capacity have been the largest with the USDA Forest Service research branch, but extend to the forest products sector and most state forestry research organizations as well.

Universities have maintained core strength in terms of the number of forestry professors, and have added many long-term temporary Ph.D. level professors and professionals as well, which are not recorded in the available data. In addition, many broader ecological and social faculty and programs in broader natural resources departments and colleges now contribute in part to forestry research and education. Recent budget cuts in all states in the 2000s, however, suggest that stable academic research and education support is likely to be at risk as well now. New research funds such as those from DOE (pre-competitive, productivity), USDA (NRI basic biology), NASA (remote sensing and GIS), and EPA (water and air quality, pollutants, and mitigation) have contributed the largest increases in funds and scope of forestry research in the last decade.

This mix of reductions in the research and education capacity of traditional Forest Service and forest industry organizations, stable levels (but dynamic fluctuations) in universities, and growth in new areas of forestry research makes universal generalizations difficult. The new research areas provide an example of enhanced prospects for forestry research capacity, but are neither comprehensive nor adequate by themselves. Systematic, broad-based, and thoughtfully planned programs as well as strengthened resources are required if forestry research capacity will meet rapidly increasing demands for a wealth of goods and services.

In order to revitalize our forestry research and education capacity, especially in the traditional agencies and organizations, this report makes eleven principal recommendations. They are discussed again briefly below in order of presentation in the report.

Recommendation 2-1
To achieve an adequate knowledge base, forestry and natural-resource education and research programs in government and academia should dedicate resources to the foundation fields of forestry science while engaging in efforts to develop emerging education and research priority areas.

Forestry research and education opportunities can be viewed as a continuum spanning foundation and emerging disciplines. The foundation disciplines are based in the traditional areas of biology and management that have been a focus since forestry began. They remain crucial in determining how forest resources are classified, managed, and protected. Allocation of resources to these traditional programs, which currently appear to be inadequately addressed, is key and renewed student interest is at least one prerequisite for foundation topics to prosper.

It is clear, however, that many basic scientific and educational questions still rest on the foundation disciplines. Thus their principles must be taught even as students and funding migrates toward more contemporary subjects such as ecology, remote sensing and GIS, human dimensions, or biotechnology. Furthermore, these emerging disciplines require resources to support the keen interest demonstrated by current forestry and natural resources students while maintaining efforts to improve our knowledge in the fundamental disciplines.

Recommendation 3-1
The Forest Service should enhance its current research-information system and tracking efforts by establishing an improved and integrated interagency system that includes relevant information on forestry research activities, workforce, funding, and accomplishments in all agencies of the U.S. Department of Agriculture, other relevant federal agencies, and associated organizations as appropriate.

One persistent challenge in preparing this report was the lack of data on current forestry research personnel, infrastructure, and support. The USDA Forest Service collects data on its scientists, Research Work Units, and budgets. The USDA has compiled statistics on forestry employment at universities periodically, and some data on student trends are collected by the Food and Agriculture Education Information System (FAEIS). Data on forestry research and education support in other agencies are scarce. To continually assess the state of forestry research, better data and coordinated efforts for collecting it are needed. This recommendation parallels the general principles of adaptive management, only extended to social institutions. It also corresponds to general efforts to track research efforts that are required by the SFM Criteria and Indicators, by government performance monitoring, and by forest certification organizations.

Recommendation 3-2
The Forest Service should substantially strengthen its research workforce over the next five years to address current and impending shortfalls, specifically recruiting and retaining researchers trained in the disciplines identified as foundation and critical emerging fields of forestry science.

A decrease of 25% in the number of Forest Service scientists over a 10-year period (1978-1988) was reported in 1990. In the subsequent 10 years (1988-1998), Forest Service scientists decreased by another 25%. In total, from 1985 to 1999, the number of Forest Service research scientists declined from 985 to 537, or by almost half. The declines in the Forest Service scientific workforce have had significant adverse impacts on the breadth and depth of the agency's research efforts. Continuation of these declines will dangerously erode the ability of the agency to answer questions in fundamental disciplines, where they once were the leaders in the world. Furthermore, the agency will be less likely to be able to contribute effectively in the emerging disciplines, which often require prompt response as well as long-term focused efforts.

Recommendation 3-3
As part of the increase in research personnel capacity and resources, the Forest Service should enhance cooperative relations with forestry schools and colleges.

As described in Chapters 3 and 5 of this report, collaborative efforts among the federal government and universities have provided very successful means of improving research and education. Such efforts allow scientists from the federal government to work closely with those at relevant universities. Cooperation allows more effective use of scarce human, financial, and equipment resources, and better exchange of new ideas to prompt innovation. The ability to cooperate at great distances has increased significantly as electronic communications have become easier. There are many examples of federal cooperative efforts, such as the U.S Geological Survey Fish and Wildlife Service Cooperative Research units, which have been models of how cooperation can be effective.
Forest Service extramural support for university research increased moderately from 1980 to 1995, but has declined precipitously since. By 1998, Forest Service extramural funding was at its lowest level in real dollars since 1980. Many of these grants are earmarked for supporting ongoing contracts and relationships that effectively help extend the agency's workforce, so actual discretionary extramural funds are even less than indicated by the data. In addition, all of the Forest Service extramural funding is allocated through negotiation with individual Research Work Units and collaborating university scientists rather than through some type of open competition. This has tended to concentrate available funds and most cooperation on a rather narrow set of traditional activities and players. The levels of interaction and cooperation between the Forest Service and university researchers must be increased to help both sectors achieve common research goals and interests. More open research grant processes will help enhance broader participation, better science, and more support as well.

Recommendation 3-4
The USDA Forest Research Advisory Committee should focus its efforts in two primary areas: (1) work with USDA research leaders in the Forest Service and other agencies to set research priorities and monitor accomplishments, and (2) coordinate with USDA's Cooperative State Research, Education, and Extension Service and other agencies to help guide research priorities of McIntire-Stennis, Renewable Resources Extension Act, National Research Initiative, and other grant programs.

Forestry research is a dynamic system that has evolved slowly to encompass a host of fundamental and emerging disciplines. Accompanying this change has been the addition of many new forestry research organizations and funding mechanisms, both within the USDA Forest Service, within USDA, and in other federal, state, or private organizations. At the same time, the scope of forestry research has broadened to encompass many natural resource disciplines, including such subjects as wildlife, biodiversity, water quality, air quality, human dimensions, regional economics, and international issues. Given these trends, better coordination, consultation, and collaboration within USDA and among other forest research entities is needed.

The existing USDA Forest Research Advisory Committee (FRAC) could serve as a means to enhance collaboration among current stakeholders and the Forest Service. The last results of a Forest Service planning exercise were published in 1990, and no formal national research vision or stakeholder consultation has occurred since. Similarly, the specific research activities funded by McIntire-Stennis and USDA NRI have had only modest program reviews. The FRAC committee has been modified in the past and could be modified in composition and in charter to provide some ongoing consultations strategic directions for the Forest Service and other USDA federal forestry or for formula funds. If possible, other federal and nongovernmental forestry research organizations should be invited to participate.

Recommendation 3-5
Universities and state institutions should increase the use of competitive mechanisms for allocating McIntire-Stennis and Renewable Resources Extension Act funds within these institutions, and in doing so, encourage team approaches to solving forestry and natural resource problems as well as integrated research and extension proposals or interinstitutional cooperation.

The university forestry research sector benefits substantially from federal formula funds through the McIntire-Stennis program, as have land grant agricultural schools and experiment stations through the Hatch Act. These funds provide base support for forestry (or agriculture) research, supplementing state appropriations and external fund sources. McIntire-Stennis formula funds are distributed among all states, ensuring that even small forestry programs receive some funds and can perform some applied research in their state. This promotes equity among states, which all have some forest resources, as well as broad-based political support.

In the last decade competitive grants have been recognized and widely supported by scientific organizations (National Research Council 1989, 1990, 1994, 1996, 2000).

Most new grants in forestry, such as those provided by NRI, NSF, DOE, and EPA are administered by open, competitive requests for proposals.

Overall, the need for some base support of funding for forestry research and education across all states and disciplines, and the merits of having a mix of funding sources, suggests the importance of McIntire-Stennis funding. In order to enhance the merits of formula funding, its allocation within states or at specific institutions could be enhanced. Possible improvements might be more focus on interdisciplinary research; joint efforts among states or scientists; better integration of extension components into the research and student education functions; and more competition for funds at a given institution.

Recommendation 3-6
The U.S. Department of Agriculture, together with universities, should develop means to more effectively communicate existing and new knowledge to users, managers, and planners in forestry.

Enhanced forestry outreach and extension efforts continue to be a key to successful implementation of forestry research and professional education efforts. The cooperative extension program has successfully transferred knowledge about forest productivity and protection for decades, and expanded its mission to include programs in economic development, urban forestry, environmental education, and nontimber forest products. More integrated programs of research, education, and extension must be developed to ensure that the research recommendations suggested here are carried out. Enhanced forestry extension for nonindustrial private forest landowners was one of the priority recommendations in the recent National Research Council report *Forested Landscapes in Perspective: Prospects and Opportunities for Sustainable Management of America's Nonfederal Forests* (National Research Council, 1998) and remains as salient for achieving enhanced research capacity.

Recommendation 4-1
University programs should assume a renewed commitment to the fundamental areas of scholarship and research in forest sciences that have diminished in recent years, and should adopt an enhanced, broad, integrative, and interdisciplinary programmatic approach to curricula at the graduate level.

Chapter 5 of this report notes that university graduate degrees must provide students with programs that have depth, breadth, integration, and diversity. The foundation areas of forestry research traditionally provided graduate students with great depth in a particular subject area. Students often have been encouraged to become even more specialized by the reductionist nature of modern research. Specialization is certainly accentuated in fields such as forest biotechnology, but even this smallest scale of research triggers an incredibly complex set of social, moral, and ethical questions that students must be aware of. Many of the emerging fields of research are oriented toward a systems approach to investigation and inference. Thus some disciplines such as ecology and spatial information must by nature be more interdisciplinary. The need to integrate even narrow research results into a larger research portfolio has become more important. Thus our education system must ensure that graduate students can perform reductionist science as needed to answer fundamental questions, but be able to translate and aggregate such results into the broader biophysical world and social framework. The forestry research and education sector must additionally attract a graduate student population that is more diverse than at present, and keep and promote those individuals once they become forestry research professionals.

Recommendation 4-2
Universities should develop joint programming in regions to ensure a "critical mass" of faculty and mentoring expertise in fields where expertise might be dispersed among the universities.

This recommendation stems from new needs driven by increasing fiscal and human resource constraints given increasingly scarce public and private resources for forestry research. Various forestry schools are attempting to develop new models that integrate broad themes, such as sustainable forest management and forest productivity in the South or ecosystem management in the West. The new models have involved programs in which students take courses at different universities—such as Washington State University and the University of Idaho, which are within 10 miles of each other—or development of widely needed courses by the western National Association of Professional Forestry Schools and Colleges (NAPFSC) schools. The University of Georgia and North Carolina State University have proposed development of a "virtual center" that encourages cooperation among universities. The idea is to integrate the strengths of the forestry schools and to do team research focused on high-priority needs.

Recommendation 5-1
Centers of excellence in forestry should be established and administered by USDA. These programs and awarded projects should (1) support interdisciplinary and interorganizational activities, (2) focus on increasing minority student participation in education and research, (3) clearly justify how new forestry-research approaches and capacity will be enhanced, and (4) undergo initial and periodic review.

A committee of the National Research Council called for the establishment of "centers of emphasis" for forestry research in 1990, and this concept remains relevant today. The committee concluded that there must be at least one center for each of five research subject areas they described. They also made the point that a center of excellence need not be bricks and mortar but instead could be a corporate mechanism that allows scientists to interact in a manner that enhances their productivity. Similarly, the current report discusses the value of "virtual" centers.

For existing or expanded forestry research to be more effective, means must be found to reshape organizational structures and cooperative efforts to capitalize on the creativity of independent research and the accomplishments of mission-oriented research. It must be kept in mind that current organizations reflect the purposeful selection and adaptation of forestry research to prior needs. Modifications must reflect a similar consensus about today's issues in foundation and emerging forestry research.

Ideas like the centers of excellence in forestry proposed in Chapter 4 will be required, given financial limitations that dictate that we cannot have redundancy across our university system. If "sister universities" with different strengths work together, greater capacity in education and research can be created. That will allow multiple skills to be brought together to address the complex questions in such fields as ecosystem forestry, biotechnology, and intensive forest management.

In short, no single administrative structure or organization can guarantee that the appropriate forestry-research goals are defined and that research is carried out creatively and effectively. Foundation research has traditionally been performed in forestry in the classical single-scientist or at least single-institutional approach. However, breakthroughs in basic physiology, biology, and biotechnology have required large teams of scientists, and are moving toward cooperation across institutions. Emerging research in ecosystem science and social science require broad interdisciplinary and interinstitutional teams. Identifying the mechanism and the organizational structure—as well as the funding—is essential for these new fields of research to be successful. Overall, we should examine the various kinds of research we need and the types of organizations we have. Then we should try to match the research needs with existing organizations and funding and modify the organizations, cooperation among organizations, and funding to achieve the best forestry research possible with the resources available.

Recommendation 5-2
Clear federal research facility mandates—such as long-term ecological research sites, experimental forest and natural resource areas, and watershed monitoring facilities—should receive priority for retention and enhancement, and a system of periodic review of all facilities should be implemented and maintained.

Establishing forestry research management collaborations at large spatial scales with an environmental perspective was identified as a priority by the National Research Council Committee on Forestry Research in 1990, and this concept also remains salient today. In fact, many long-term ecological (LTER) sites funded by NSF have proven successful, and several Forest Service research station strategic plans have identified the importance of long-term research and monitoring as part of their priorities. In the past, Forest Service research and monitoring sites tended to be narrowly focused on only a few components, such as silviculture or hydrology. Modern LTER sites have involved much broader multidisciplinary activities on large tracts of land. This report concurs with the merits of the long-term research sites and funding, with the added component of regular, periodic review.

CONCLUSIONS

Forestry research capacity is indeed at a crossroads, and perhaps even at risk. The same observations might be made for the forestry profession as well. There is unprecedented public pressure and demands on a declining forest resource base, at a time when public expenditures are decreasing for forestry research, professional education, public extension, forest management, and natural resource protection. Our ability to manage forests to produce more goods, provide more developed and undeveloped services, harbor great biodiversity, support community development, and protect the natural environment depends on interdisciplinary and integrative research, education, and outreach efforts.

The common mantra for responding to these conflicting pressures is that forestry professionals must work smarter, harder, and more efficiently. While this is true, it is insufficient. Success will require clear technical cooperation in performing research, which provides evidence that the forestry sector is performing research efficiently. What is necessary is a concerted, permanent cooperative effort among many stakeholders, which includes joint strategic planning and monitoring; continued support of existing organizations and fundamental and emerging research; a larger and open cooperative grants programs from the Forest Service; broader training for forestry graduate students; and an integrated research, education, and extension enterprise.

Enhancing the nation's forestry-research capacity must deal with the tangible matters of substance—funding, facilities and equipment, and personnel—and with intangible matters of perception and values—priorities, organizations, structures, and leadership. This current review of programs and accomplishments, together with input from various groups, provides some guidance in many of these areas, which enables this committee to make recommendations for securing the nation's strength and capacity in forestry research.

References

Alston, J.M. and P.G. Pardey. 1996. Making Science Pay: The Economics of Agricultural Research and Development Policy. Washington, D.C.: American Enterprise Institute.

American Forest & Paper Association. 2000. Sustainable Forestry Initiative objectives and indicators. Washington, D.C.

American Forest & Paper Association. 1999. Agenda 2020: The path forward: an implementation plan. Washington, D.C.: American Forest & Paper Association. 31 p.

American Forest & Paper Association. North Central Region. 1996. Priority research needs from a forest industry and research funding point of view. Washington, D.C.: American Forest & Paper Association North Central Forest Resources Research Committee. 14 P. + appen. mimeo.

American Forest Congress. 1996. Research needed, by regions of the United States, to sustain the nation's forests into the twenty-first century. Seventh American Forest Congress. Available from: Peter Roussopoulos, Director, USDA Forest Service, Southern Research Station, Asheville, NC. 72 p. mimeo

American Forest Congress, Lake States Region. 1996. Lake states region forest research report. Available from: Dr. Alan R. Ek, Department of Forest Resources, University of Minnesota, St. Paul, MN. 24 p. mimeo.

American Forest Congress, Northeast Region. 1996. Northeast region forest research report. Available from: Dr. Ross Whaley, College of Environmental Science and Forestry, Syracuse, New York. 18 p. mimeo.

American Forest Congress, Northern Pacific Coastal Region. 1996. Northern pacific coastal region forest research report. Available from: James R. Anderson, Northwest Indian Fisheries Commission, Olympia, WA. 29 p. mimeo.

American Forest Congress, Southern Region. 1996. Southern region forest research report. Available from: Peter Roussopoulos, Director, USDA Forest Service, Southern Research Station, Asheville, NC. 22 p. mimeo.

American Forest Congress, Southwest Region. 1996. Southwest region forest research report. Available from: John A. Helms, Department of Environmental Science, Policy, and Management, University of California at Berkeley, Berkeley, CA. 13 p. mimeo.

REFERENCES

Araji, A.A. 1981. The economic impact of investment in integrated pest management. Research Bulletin 115. Moscow, ID: University of Idaho, Agricultural Experiment Station. 27 p.

Bare, B.B. and R. Loveless. 1985. An overview of the regional forest nutrition research project. In: Risbrudt, C.D. and P.J. Jakes (eds.), Forestry Research Evaluation: Current Progress, Future Directions. Proceedings of the Forestry Research Evaluation Workshop, St. Paul, MN, 20–21 August 1984. GTR–NC–104, USDA Forest Service, North Central Forest Experiment Station, St. Paul, MN. 140 p.

Baughman, William D. and R. Scott Wallinger. 1999. The employer's perspective on new hires. *Journal of Forestry* 97(9): 12–16.

Bengston, David N. 1984. Economic impacts of structural particleboard research. *Forest Science* 30(3):685–697.

Bengston, David N. 1985. Aggregate returns to lumber and wood products research: An index number approach. Pages 62–68 in: Forestry Research Evaluation: Current Progress, Future Directions, C.D. Risbrudt and P.J. Jakes (eds.). General Technical Report NC–104. St. Paul, MN: USDA Forest Service, North Central Forest Experiment Station. 140 p.

Bengston, David N. 1999. Summary of economic evaluations of forestry research. Personal communication. October 1999.

Bengston, David N. and Hans M. Gregersen. 1988. What influences forestry research capacity in developed and less–developed countries? Journal of Forestry 86(2):41–43.

Bengston, David N. and ??? Xu. 1993. Returns to recreation research.

Bentley, W.R. 1999. Professional forestry education in New York: An old lesson, a new model. Journal of Forestry 97: 29–32.

Birch, Thomas. 1996. Private Forest–land owners of the United States, 1994. Resource Bulletin NE–134. Radnor, PA: USDA Forest Service, Northeastern Forest Experiment Station. 183 p.

Brunner, Allan D. and Jack K. Strauss. 1987. The social returns to public R&D in the U.S. wood preserving industry. SCFER Working Paper No. 35. Research Triangle Park, NC: Southeastern Center for Forest Economics Research. 36 p.

Burkhart, Harold. 1999. Observations of the future of southern forest research. Speech presented at 1999 Southern Industrial Forestry Research Council Meeting on Southern Forest Industry Research Cooperatives. November 16–17. Atlanta, GA.

Callaham, R.Z. 1989. Training and education for management of RD and A. In A.L. Lundgren (ed) The management of large-scale forestry research programs and projects. USDA General Technical Report NE – 1 30. Broomall, PA, USDA Forest Science Northwestern Forest Experiment Station.

Cantrell, Richard. 1999. Sustainable Forestry Initiative summary for forest products industry research expenditures. Personal communication. October 1999.

Chang, S. Joseph. 1985. The economics of optimal stand growth and yield information gathering. Final Report, Cooperative Research Agreement 23-83-27. St. Paul, MN: USDA Forest Service, North Central Forest Experiment Station. 34 p.

Chubin, D.E. and E.J. Hackett. 1990. Peerless Science: Peer Review and U.S. Science Policy. Albany: State University of New York Press.

Committee of Scientists. March 15, 1999. Sustaining the People's Lands: Recommendations for Stewardship of the National Forests and Grasslands into the Next Century. U.S. Department of Agriculture. Washington, D.C.

Cubbage, Frederick W, Laurence Jervis, and P. Gregory Smith. 1999. Forestry employment: national perspectives, southern trends. Journal of Forestry 97(9):22–28.

Cubbage, Frederick W., et al. 1988. Impact of new technology on timber harvesting costs: Evaluation methods and literature. Paper no. 88–5031. St. Joseph, MI: American Society of Agricultural Engineers. 13 p.

Cubbage, Frederick W., John M. Pye, Thomas P. Holmes, and John E. Wagner. 2000. An economic evaluation of fusiform rust protection research. Southern Journal of Applied Forestry. Accepted for publication.

Ellefson, Paul V. and Alan R. Ek. 1996. Privately initiated forestry and forest products research and development: Current Status and future challenges. Forest Products Journal 46(2):37–43.

Food and Agricultural Education Information System. 1997a. Fall 1996 Enrollment for Agriculture, Renewable Natural Resources and Forestry. Texas A&M University, College Station, Texas.

Food and Agricultural Education Information System. 1997b. Degrees Awarded and Placement for Agriculture, Renewable Resources and Forestry: Academic Year 1995/1996. Texas A&M University, College Station, Texas.

Food and Agricultural Education Information System. 1998a. Fall 1997 Enrollment for Agriculture, Renewable Natural Resources and Forestry. Texas A&M University, College Station, Texas.

Food and Agricultural Education Information System. 1998b. Degrees Awarded and Placement for Agriculture, Renewable Resources and Forestry: Academic Year 1996/1997. Texas A&M University, College Station, Texas.

Food and Agricultural Education Information System. 1999a. Fall 1998 Enrollment for Agriculture, Renewable Natural Resources and Forestry. Texas A&M University, College Station, Texas.

Food and Agricultural Education Information System. 1999b. Degrees Awarded and Placement for Agriculture, Renewable Resources and Forestry: Academic Year 1997/1998. Texas A&M University, College Station, Texas.

Food and Agriculture Organization. July 1999. State of the World's Forests, 1999. Available on–line at: http://www.fao.org/fo/sofo/sofo99/pdf/sofe_e/coper_en.pdf.. Rome: Food and Agriculture Organization of the United Nations.

Food and Agriculture Organization. 2001. State of the World's Forests, 2001. Rome: Food and Agriculture Organization of the United Nations. 181 p.

Food and Agricultural Organization. February 2002. Definitions and Examples of Terms Commonly Used to Describe What Forests Provide. Available on-line at: http://www.fao.org/docrep/V7540e/V7540e28.htm.

Forest Stewardship Council. 2001. Forest Stewardship Council principles and criteria. Document 1.2, Revised February 2000. http//www/fscoax.org/html.noframes/1-2.html.

REFERENCES

Fox. B.E., T.E. Kolb, and E.A. Kurmes. 1996. An integrated forestry curriculum: The Northern Arizona University experience. Journal of Forestry 94: 16–22.

Ginger, C., D. Wang, and L. Tritton. 1999. Integrating disciplines in an undergraduate curriculum. Journal of Forestry 97: 17–21.

Gordon, J.C. 1984. Educating Tomorrow's Foresters. American Forests 90(4): 10–56.

Guldin, Richard. 1999. Forestry research statistics for the USDA Forest Service; budget trends, research locations and Research Work Units, and Publication productivity. Personal communication. July 1999.

Guldin, Richard. 1999. Forestry research statistics for the USDA Forest Service; Forest Service funding by Budget Line Item, and trends in employment in universities. Personal communication. October 1999.

Haygreen, John, et al. 1986. The economic impact of timber utilization research. Forest Products Journal 36(2):12–20.

Helms, John, Editor. 1998. Dictionary of Forestry. Washington, D.C.: Society of American Foresters. 210 p.

Huang, Y. Star, and Laurence Teeter. 1990. An economic evaluation of research on herbaceous weed control in southern pine plantations. Forest Science 36(2): 313–329.

Huffman, W.E. and R.E. Evenson. 1993. Science for Agriculture: A Long-Term Perspective. Ames: Iowa State University Press.

Hyde, William F., David H. Newman, and Barry J. Seldon. 1992. The Economic Benefits of Forestry Research. Ames, Iowa: Iowa State Univ. Press. 249 p.

Levenson, Burton E. 1984. Economic analysis of tree improvement research in Michigan. Ph.D. dissertation, Michigan State University, East Lansing, MI. 176 p.

Lewis, R., Deputy Chief Forest Service Research and Development. Forest Service Research Oversight. March 28, 2000. United States Department of Agriculture. House of Representatives Committee on Appropriations Subcommittee on Interior and Related Agencies.

Little, A.D., Inc. 1991. Third Generation R&D: Managing the Link to Corporate Strategy. P. Roussel, K. Saad, and T. Jackson.

Lucier, Alan. 1999. Summary of NCASI forestry research budgets. Personal communication. November 1999.

Meridian Institute. March 2002. Comparative Analysis of the Forest Stewardship Council and Sustainable Forestry Initiative Certification Programs. Available on-line at: http://www.merid.org/comparison.

Moore, Allen. National Summary of Forestry Research Programs for Non-Federal Locations: Fiscal Year 2000 Funds and Scientist Years (Forestry Adjusted). Personal communication. Washington, DC: U.S. Department of Agriculture Cooperative State Research, Education, and Extension Service.

Munson, Ken. 1999. Observations of the future of southern forest research. Panel comments presented at 1999 Southern Industrial Forestry Research Council Meeting on Southern Forest Industry Research Cooperatives. November 16–17. Atlanta, GA.

National Association of State Foresters. 1997. Forests for a sustainable future: the use of criteria and indicators in sustainable forest management. White Paper, September 1997. 9 p.

National Association of Professional Forestry Schools and Colleges. 1999. The role of research, education and extension in sustaining America's forest resources: why you should care. Washington, D.C.: National Association of Professional Forestry Schools and Colleges and Cooperative State Research, Education, and Extension Service. 15 p.

National Academy of Sciences. 1967. Undergraduate Education in Renewable Natural Resources: An Assessment. Washington, D.C.: Printing and Publishing Office.

National Research Council. 1926. Forest Research under State Auspices by F.W. Besley April, 13 pp.

National Research Council. 1927a. Research in Biological Sciences Fundamental to Forestry. Committee on Forestry Research. Subcommittee on Survey of Status of Research in Forestry.

National Research Council. 1927b. Education in Forestry. Committee on Forestry Research. Subcommittee on Survey of Status of Research in Forestry.

National Research Council. 1928. Forestry Research in the United States.

National Research Council. 1947. Problems and Progress of Forestry in the United States. Joint Committee on Forestry of the National Resource Council and the Society of American Foresters; report of the Joint Committee on Forestry of the National Research Council and the Society of American Foresters, Henry S. Graves, chairman, Washington, D.C. Society of American Foresters.

National Research Council. 1989. A Proposal to Strengthen the Agricultural, Food, and Environmental System. Washington, D.C.: National Academy Press. 155 p.

National Research Council. 1990. Forestry Research: A Mandate for Change. Washington, D.C.: National Academy Press. 84 p.

National Research Council. 1994. Investing in the National Research Initiative: An Update of the Competitive Grants Program in the U.S. Department of Agriculture. Washington, D.C.: National Academy Press. 64 p.

National Research Council. 1995. Reshaping the Graduate Education of Scientists and Engineers. Washington, D.C.: National Academy Press, 208 p.

National Research Council. 1996. Colleges of Agriculture at the Land Grant Universities: Public Service and Public Policy. Washington, D.C.: National Academy Press., 121 p.

National Research Council. 1998. Forested Landscapes in Perspective: Prospects and Opportunities for Sustainable Management of America's Nonfederal Forests. Washington, D.C.: National Academy Press. 249 p.

National Research Council. 1999. Evaluating Federal Research Programs: Research and the Government Performance and Results Act. Committee on Science, Engineering, and Public Policy. Washington, D.C.: National Academy Press. 80 p.

National Research Council. 2000. National Research Initiative: A Vital Competitive Grants Program in Food, Fiber, and Natural-Resources Research. Washington, D.C.: National Academy Press. 189 p.

National Science and Technology Council. 1997. Integrating the nation's environmental monitoring and research networks and programs: a proposed framework. Environmental Monitoring Team. Committee on Environment and Natural Resources, March 1997, 102 p.

National Science and Technology Council. 1999. Renewing the Federal Government-University Research Partnership for the 21st Century. Committee on Science. Executive Office of the President. Office of Science and Technology Policy. 13 p.

National Science and Technology Council. 1999. Improving Federal Laboratories to Meet the Challenges of the 21st Century. Interagency Working Group on Federal Laboratory Reform. Office of Science and Technology Policy, 26 p.

National Science and Technology Council. 2001. Implementation of the National Science and Technology Council Presidential Review Directive 4: Renewing the Federal Government-University Research Partnership for the 21st Century. January 2001, 19 p.

National Science Foundation. 1998. Science and Engineering Indicators. National Science Board, National Science Foundation, Arlington, VA, 32 p.

National Science Foundation. 1996. Academic Research Instruments: Expenditures 1993, Needs 1994. NSF 96-324, Arlington, VA, 46 p..

Noble, Ian R. and Rodolfo Dirzon. 1997. Forests as human–dominated ecosystems. Science 277(5325):522–525.

O'Laughlin, Jay, Lita C. Rule, and Christopher D. Risbrudt. 1986. U.S. Forest Service Research: What proportion sustains (vs. enhances) productivity? In: Proceedings, of the 1986 Southern Forest Economics Workshop. Available from Dr. D.L. Holley, NC State University, Box 8002, Raleigh, NC 27695. p. 117–127.

Powell, Douglas S., Joanne L. Faulkner, David R. Darr, Zhiliang Zhu, and Douglas MacCleery. 1993. Forest Resources of the United States, 1992. General technical report RM–234. Fort Collins, CO: USDA Forest Service, Rocky Mountain Range and Experiment Station. 132 p. + map. [Revised, June 1994].

Pye, John M., John E. Wagner, Thomas P. Holmes, and Frederick W. Cubbage. 1997. Positive returns from investment in fusiform rust research. Research Report SRS–4. USDA Forest Service, Southern Research Station. 55 p.

Roundtable on Sustainable Forests. March 2002. Available on-line at: http://www.sustainableforests.net/.

Roussopolous, Peter. 1999. Observations of the future of southern forest research. Panel comments presented at 1999 Southern Industrial Forestry Research Council Meeting on Southern Forest Industry Research Cooperatives. November 16–17. Atlanta, GA.

Sample, V.A., P.C. Rinegold, N.E. Block, and J.W. Giltmier. 1999. Forestry education: Adapting to the changing demands. J. Forestry 97(9):4-10.

Seldon, Barry J. and David H. Newman. 1987. Marginal productivity of public research in the softwood plywood industry: A dual approach. Forest Science 33(4):872–888.

Siry, Jacek P, Frederick W. Cubbage, and Andy Malmquist. 2001. Potential impacts of increased management intensities on planted pine growth and yield and timber supply modeling in the South. Forest Products Journal 51(3):42-48.

Smith, W., J. Brad, L. Visage, R. Sheffield, and D.R. Darr. 2001. Forest resources of the United States, 1997. General Technical Report NC-000. USDA Forest Service North Central Forest Experiment Station, St. Paul, MN. January 2001.

Society of American Foresters. 1998. Guidelines for Professional Forestry Accreditation. Bethesda, MD: Society of American Foresters.

Society of American Foresters. 1999. Proposed New Guidelines for Professional Forestry Accreditation; Unpublished Committee Report. Bethesda, MD: Society of American Foresters.

Southern Industrial Forestry Research Council (SIFRC). 1996. Priority research needs from a forest industry view. Southern Industrial Forestry Research Council Report No. 6. Washington, D.C.: American Forest & Paper Association.

Stoltenberg, C.H., Ware, K.D., Marty, R.J., Wray, R.D., and Wellons, J.D. 1970. Planning Research for Resource Decisions. Iowa State University Press. Ames, IA. 183 pp.

Szaro, R.C., A.M. Yapi, D. Langor, E. Schaitza, K. Awang, and K. Vancura. 2000. Forest Science Challenges and Contributions to Sustainable Human and Resource Development. In Forest Science Challenges and Contributions.

U.S. Department of Agriculture. 1985. Directory of Professional Workers in State Agricultural Experiment Stations and Other Cooperating State Institutions. Washington, D.C.: U.S. Department of Agriculture's Cooperative States Research Service Handbook No. 305.

U.S. Department of Agriculture. 1987. Directory of Professional Workers in State Agricultural Experiment Stations and Other Cooperating State Institutions. Washington, D.C.: U.S. Department of Agriculture's Cooperative States Research Service Handbook No. 305.

U.S. Department of Agriculture. 1990. Strategy for the 90's for USDA Forest Service Research. Washington, D.C.: USDA Forest Service.

U.S. Department of Agriculture. 1993. The Principal Laws Relating to Forest Service Activities. Washington, D.C.: U.S. Government Printing Office.

U.S. Department of Agriculture. 1994. Directory of Professional Workers in State Agricultural Experiment Stations and Other Cooperating State Institutions. Washington, D.C.: U.S. Department of Agriculture's Cooperative States Research Service Handbook No. 305.

U.S. Department of Agriculture. 1998. Reports of the Forest Service: Fiscal Year 1980–98. Washington, D.C.: USDA Forest Service.

U.S. Department of Agriculture. 1999. Report of the Forest Service: Fiscal Year 1998. Washington, D.C.: USDA Forest Service.

REFERENCES

U.S. Department of Agriculture. 1999. Report of the Strategic Planning Task Force on USDA Research Facilities: Report and Recommendations. Washington, D.C. USDA. 108 p.

U.S. Department of Agriculture. 2002. A Review of the Capacity to Conduct and Apply Research and Development Aimed at Improving Forest Management and Delivery of Forest Goods and Services. Review draft, 7 March 2002. Washington, D.C.: USDA Forest Service.

U.S. Department of Labor. 2000. Workforce 2000.

University of Idaho. 1983. Our Natural Resources: Basic Research Needs in Forestry and Renewable Natural Resources. National Task Force on Basic Research in Forestry and Renewable Resources. The Forest, Wildlife, and Range Experiment Station, Moscow, 35p.

Wallinger, R.S. 1991. Creating and educating a 21st century forest resources professional. Pp. 30–38 in Forest Resources Management in the 21st Century: Will Forestry Education Meet the Challenge? Conference Proceedings. Society of American Foresters. Bethesda, MD.

Westgate, Robert A. 1986. The economics of containerized forest tree seedling research in the United States. Canadian Journal of Forest Research 16:1007–1012.

Whaley, Ross. 1999. Reflections of forestry research structure and organizations. Paper presented at: National Research Council Meeting on National Capacity in Forestry Research. July 15–16, 1999. Washington, D.C.

Williams, Michael. 1994. Forests and tree cover. p. 97–124. In: William B. Meyer and B.L. Turner II (editors): Changes in Land use and Land Cover: A Global Perspective. Cambridge, England: Cambridge University Press.

World Commission on Environment and Development. 1987. Our Common Future. Oxford: Oxford University.

World Resources Institute et al. 1998. World Resources 1998–99. Joint publication by the World Resources Institute, The United Nations Environment Programme, The United Nations Development Programme, and The World Bank. New York: Oxford University Press.

Yin, Runsheng, Leon V. Pienaar, and Mary Ellen Aronow. 1998. The productivity and profitability of fiber farming. Journal of Forestry 96(11):13–18.

Appendix A

AGENDA

Workshop on National Capacity in Forestry Research

July 15 & 16, 1999

Lecture Room
National Academy of Sciences
2101 Constitution Avenue, NW
Washington, DC 20418

Thursday, July 15, 1999

8:15 AM	Welcome and Introductory Remarks *Frederick W. Cubbage, Committee Chair*
8:30	Perspectives on Future Research Needs *Barbara C. Weber, USDA Forest Service, Washington, DC*
9:00	*Assessment of Current Research Capacity* John Pait, The Timber Company, Atlanta, GA
9:30	*Merits and Limits of Current Approaches and Structures* Ross Whaley, State University of New York, College of Environmental Science and Forestry
10:00–10:30	Break
10:30	<u>Panel: Research Needs and Opportunities</u> *Moderator: Arnett C. Mace, Jr., University of Georgia* Biological Diversity and Sustainability *Norman Christensen, Duke University, Durham, NC* Forestry Research Conservation and Reserves *Christopher Haney, Wilderness Society, Washington, DC* Research Needs and Opportunities for Forest Landowners *Kirk Rodgers, Forest Landowners Association, Washington, DC* Intensive Production and Forest Zonation *Clark Binkley, Hancock Timber Resource Group, Boston, MA*

APPENDIX A

                Biotechnology and Genomics, Biological and Fundamental Sciences
                Les Pearson, Westvaco, Summerville, SC

                Global Competition and Wood as Raw Material
                James Bowyer, University of Minnesota

11:30   General Discussion

12:00–1:00 PM Break

1:00 PM   <u>Panel: Research Responses to Forestry Needs</u>
                Moderator: Thomas J. Mills, Pacific Northwest Research Station, Portland, OR

                Development and Application of Interdisciplinary Research
                Bernard Bormann, USDA Forest Service, Corvallis, OR

                Forest Management for Ecological Benefits and Species Protection
                Danna Smith, Dogwood Alliance, Brevard, NC

                Community Based Conservation, Management and Economic Development
                Jonathan Kusel, Forest Community Research, Taylorsville, CA

                Forestry Graduate Curricula Directions
                Alan Ek, University of Minnesota

                Urban Forestry
                Gerald Gray, American Forests, Washington, DC

2:00   General Discussion

2:45   Breakout Session I
                Attendees divided into working groups. A leader and recorder appointed for each group. Groups were asked to identify critical forestry issues and priorities for forestry research.

4:30   Reports from Breakout Session Leaders

5:00   Adjourn

Friday, July 16, 1999

8:00 AM   Review of the Previous Day's Activities
                Frederick Cubbage

8:15   Breakout Session II: Examining Hypotheses

Attendees divided into working groups. A leader and a recorder appointed for each group. Groups were asked to examine hypotheses related to meeting forestry research priorities identified during Breakout Session I.

10:00	Break
10:15	Reports from Breakout Session Leaders
10:45	Comments and Input from the Public
11:45	Conclusions *Frederick W. Cubbage, Committee Chair*
12:00	Adjourn

Appendix B

BREAKOUT GROUP QUESTIONS

Workshop on National Capacity in Forestry Research

July 15 & 16, 1999

Lecture Room
National Academy of Sciences
2101 Constitution Avenue, NW
Washington, DC 20418

Breakout Session I
July 15, 1999

1) Blue Group: What are critical forestry issues and priorities for forestry research?
2) Green Group: What are critical forestry issues and priorities for forestry research?
3) Red Group: What are critical forestry issues and priorities for forestry research?
4) Yellow Group: What are critical forestry issues and priorities for forestry research?

Breakout Session II
July 16, 1999

1) Blue Group: Is there an adequate knowledge base?
—Identify major gaps in the knowledge base needed for forestry research.

2) Green Group: Is there adequate research capacity?
—Identify strengths and weaknesses of current research capacity

3) Red Group: Are there adequate interdisciplinary and scale applications?
—Identify needed interdisciplinary and spatial applications and incentives.

4) Yellow Group: Are university curricula and programs adequate?
—Identify strengths and weaknesses of university curricula and programs to provide researchers.

ABOUT THE AUTHORS

FREDERICK W. CUBBAGE is professor and head of Department of Forestry, North Carolina State University. He has significant expertise in forest resource policy and law, and an extensive background in forest production and harvesting economics. Before joining North Carolina State University, Dr. Cubbage served as a project leader for the Economics of Forest Protection and Management Work Unit of the Southeastern Forest Experiment Station in Research Triangle Park. He also was professor at School of Forest Resources, University of Georgia, and prior to that, he served as an associate economist at Southern Forest Experiment Station in New Orleans. In 1996, Dr. Cubbage served on a steering committee for the NRC's study on Wood as Raw Material for Industrial Use. Currently, Dr. Cubbage serves on the editorial board of the new international journal of Forest Policy and Economics, and is a working group officer for several national and international professional and research organizations. Dr. Cubbage is an author or co-author of more than 300 publications and senior author of a textbook on *Forest Resource Policy*. His professional recognitions include: Who's Who in Science and Engineering, Who's Who in the South and Southwest, and Who's Who in the World. He received the Distinguished Science and Research Award from the Southeastern Society of American Foresters (1989). He also is a member of the Forest Economics and Policy Program Advisory Board of Resources for the Future. Dr. Cubbage received his Ph.D. in forest economics from the University of Minnesota in 1981.

PERRY J. BROWN has significant expertise in natural resource policy and planning, and in recreation behavior and planning. Dr. Brown is dean and professor, School of Forestry, University of Montana. He also is the director of the Montana Forest and Conservation Experiment Station. In the past, Dr. Brown has held formal advisory appointments with the USDA Forest Service and the USDI Bureau of Land Management. Prior to his current positions, Dr. Brown served as associate dean of the College of Forestry and professor of forest recreation and resource planning at Oregon State University. While at Oregon State University, Dr. Brown also served as Social Science Project leader; interim director of Oregon State System of Higher Education Oregon Tourism Institute; director of International Programs, College of Forestry; and head and professor of Resource Recreation Management. Dr. Brown has taught numerous short courses and workshops in over 30 countries. In 1966 he was elected to Xi Sigma Pi. Dr. Brown's professional recognitions include the USDA Forest Service Chief's Certificate of Appreciation (1988) and Utah State University's Professional Achievement Award (1996). Dr. Brown received his Ph.D. in outdoor recreation and social psychology from Utah State University in 1971.

THOMAS R. CROW was Theodore Roosevelt Professor of Ecosystem Management in the School of Natural Resources and Environment, University of Michigan while serving on the committee. He has extensive expertise in managing a broad program of integrated research that includes biologic and social sciences at the University of Michigan. Prior to his position with the University of Michigan, Dr. Crow was a research ecologist with the USDA Forest Service and worked on technologies for managing temperate forests. He received his Ph.D. in forest ecology from University of Minnesota in 1970 and his M.F. in forest biology from the University of Michigan in 1965.

JOHN C. GORDON has long-standing experience in forestry education programs. He has expertise in forestry research, with emphasis on photosynthesis and translocation in trees, enzymes in woody plants, and nitrogen fixation. Dr. Gordon is Pinchot professor and former dean of the School of Forestry and Environmental Studies at Yale University. He served as chair of the NRC Committee on Forestry Research, and as a member of the NRC's Committee on Land Grant Colleges of Agriculture. Dr. Gordon is a member of the AAAS, Phi Kappa Phi, Sigma Xi, and the Society of American Foresters. He received his undergraduate and Ph.D. degrees from Iowa State University.

JOHN W. HUMKE has extensive expertise in natural resource policy. He is a vice president of The Nature Conservancy and Director of Agency Relations, operating out of the Conservancy's offices in Boulder, Colorado. He received his B.S. degree in resource management and biology from the University of Wisconsin, Steven Point, and his M.S. degree in resource development and urban planning from Michigan State University. Mr. Humke has served as the Conservancy's Midwest Regional Director and national Director of Stewardship. He currently focuses his efforts on ecoregional conservation planning and biodiversity conservation on public lands. He served as chair of the review committee for the Review of Research Natural Area Establishment Process appointed by the USDA Forest Service, and is the past president of the Natural Areas Association.

REX B. MCCULLOUGH is the vice president of timberlands, Forestry Research, at Weyerhaeuser Co., in Federal Way, WA. He joined the company in 1976 as project leader in quantitative genetics for Western Forestry in Centralia, Washington. Through his career at Weyerhaeuser Co., Dr. McCullough held several positions, including section manager for tree improvement research in Centralia; department manager of tree improvement and forestry research in Hot Springs, AR; department manager of Southern Forestry Research; director of strategic biological sciences and forest resource in Federal Way, WA; and director of timberlands, Forest Resources Research and Development. Prior to joining Weyerhaeuser Co., Dr. McCullough worked for Crown Zellerbach, Texas

A&M University, Texas Forest Service, Oklahoma State University's Forestry Department, JARI National Bulk Carriers in Brazil, Oklahoma State University's Agricultural Experiment Station Network, and the USDA's Forest Service Institute of Forest Genetics. He earned a B.S. degree in forest management in 1968 and his M.S. degree in forest genetics in 1972 from Oklahoma State University. He received his Ph.D. in forest genetics from Texas A&M University in 1975. Dr. McCullough is chair of the American Forest and Paper Association's (AF&PA) Forest Science and Technology Committee, member of the AF&PA's Agenda 2020 Committee, and the Society of American Foresters. He was co-chair of the Seventh American Forest Congress Board of Directors.

RONALD R. SEDEROFF is Distinguished University Professor of Forestry; Edwin F. Conger Professor of Forestry; and the director of the Forest Biotechnology Group in the Department of Forestry, North Carolina State University. He has extensive experience in forestry research issues. Dr. Sederoff served on the NRC committee that produced the report, Forestry Research: A Mandate for Change (1990). He also served on the NRC's Commission on Life Sciences, Board on Biology, and the Committee on Evaluation of the USDA's National Research Initiative Competitive Grants Program. Dr. Sederoff is the leader in molecular genetics of forest trees, having developed methods for genomic mapping of individual trees, complex trait analysis, and studied the molecular basis of wood properties. He received his B.A, M.A., and Ph.D. degrees in zoology (genetics) from the University of California, Los Angeles. Dr. Sederoff was elected to the National Academy of Sciences in 1995.

Board on Agriculture and Natural Resources Publications

Policy and Resources

Agricultural Biotechnology: Strategies for National Competitiveness (1987)
Agriculture and the Undergraduate: Proceedings (1992)
Agriculture's Role in K-12 Education: A Forum on the National Science Education Standards (1998)
Alternative Agriculture (1989)
Brucellosis in the Greater Yellowstone Area (1998)
Colleges of Agriculture at the Land Grant Universities: Public Service and Public Policy (1996)
Colleges of Agriculture at the Land Grant Universities: A Profile (1995)
Designing an Agricultural Genome Program (1998)
Designing Foods: Animal Product Options in the Marketplace (1988)
Ecological Monitoring of Genetically Modified Crops (2001)
Ecologically Based Pest Management: New Solutions for a New Century (1996)
Ensuring Safe Food: From Production to Consumption (1998)
Environmental Effects of Transgenic Plants: The Scope and Adequacy of Regulation (2002)
Forested Landscapes in Perspective: Prospects and Opportunities for Sustainable Management of America's Nonfederal Forests (1997)
Future Role of Pesticides for U.S. Agriculture (2000)
Genetic Engineering of Plants: Agricultural Research Opportunities and Policy Concerns (1984)
Genetically Modified Pest-Protected Plants: Science and Regulation (2000)
Incorporating Science, Economics, and Sociology in Developing Sanitary and Phytosanitary Standards in International Trade (2000)
Investing in Research: A Proposal to Strengthen the Agricultural, Food, and Environmental System (1989)
Investing in the National Research Initiative: An Update of the Competitive Grants Program in the U.S. Department of Agriculture (1994)
Managing Global Genetic Resources: Agricultural Crop Issues and Policies (1993)
Managing Global Genetic Resources: Forest Trees (1991)
Managing Global Genetic Resources: Livestock (1993)
Managing Global Genetic Resources: The U.S. National Plant Germplasm System (1991)

National Research Initiative: A Vital Competitive Grants Program in Food, Fiber, and Natural-Resources Research (2000)
New Directions for Biosciences Research in Agriculture: High-Reward Opportunities (1985)
Pesticide Resistance: Strategies and Tactics for Management (1986)
Pesticides and Groundwater Quality: Issues and Problems in Four States (1986)
Pesticides in the Diets of Infants and Children (1993)
Precision Agriculture in the 21st Century: Geospatial and Information Technologies in Crop Management (1997)
Professional Societies and Ecologically Based Pest Management (2000)
Publicly Funded Agricultural Research and the Changing Structure of U.S. Agriculture (2002)
Rangeland Health: New Methods to Classify, Inventory, and Monitor Rangelands (1994)
Regulating Pesticides in Food: The Delaney Paradox (1987)
Soil and Water Quality: An Agenda for Agriculture (1993)
Soil Conservation: Assessing the National Resources Inventory, Volume 1 (1986); Volume 2 (1986)
Sustainable Agriculture and the Environment in the Humid Tropics (1993)
Sustainable Agriculture Research and Education in the Field: A Proceedings (1991)
Toward Sustainability: A Plan for Collaborative Research on Agriculture and Natural Resource Management (1991)
Understanding Agriculture: New Directions for Education (1988)
Use of Drugs in Food Animals: Benefits and Risks, The (1999)
Water Transfers in the West: Efficiency, Equity, and the Environment (1992)
Wood in Our Future: The Role of Life Cycle Analysis (1997)

Nutrient Requirements of Domestic Animals Series and Related Titles

Building a North American Feed Information System (1995)
Metabolic Modifiers: Effects on the Nutrient Requirements of Food-Producing Animals (1994)
Nutrient Requirements of Beef Cattle, Seventh Revised Edition, Update (2000)
Nutrient Requirements of Cats, Revised Edition (1986)
Nutrient Requirements of Dairy Cattle, Seventh Revised Edition (2001)
Nutrient Requirements of Dogs, Revised Edition (1985)
Nutrient Requirements of Fish (1993)
Nutrient Requirements of Horses, Fifth Revised Edition (1989)
Nutrient Requirements of Laboratory Animals, Fourth Revised Edition (1995)
Nutrient Requirements of Poultry, Ninth Revised Edition (1994)
Nutrient Requirements of Sheep, Sixth Revised Edition (1985)
Nutrient Requirements of Swine, Tenth Revised Edition (1998)
Predicting Feed Intake of Food-Producing Animals (1986)
Role of Chromium in Animal Nutrition (1997)
Scientific Advances in Animal Nutrition: Promise for the New Century (2001)
Vitamin Tolerance of Animals (1987)

Further information, additional titles (prior to 1984), and prices are available from the National Academy Press, 2101 Constitution Avenue, NW, Washington, D.C. 20418, 202–334–3313 (information only). To order any of the titles you see above, visit the National Academy Press bookstore at http://www.nap.edu/bookstore.